Petroleum Production Operations

by
Lewis W. Hall

Edited by
Jodie Leecraft

Published by
Petroleum Extension Service
Division of Continuing Education
The University of Texas at Austin
Austin, Texas
1986

Third Impression 2013
Printed in the United States of America

Catalog No. 3.90210
ISBN 0-88698-124-7

CONTENTS

1. FUNDAMENTALS OF PETROLEUM RESERVOIRS 1
  Reservoir Rock Permeability 1
    Absolute Permeability 1
    Relative Permeability 2
  Reservoir Fluids 3
    Solubility of Gas in Crude Oil 3
    Formation Volume Factor 3
    Oil Viscosity 4
  Fluid Flow in the Reservoir 5
    Radial Flow 5
    Darcy's Law 5
2. RESERVOIR DRIVE MECHANISMS AND THEIR EFFECTS 7
  Sources of Reservoir Energy 7
  Types of Reservoir Drives 7
    Dissolved-Gas Drive 7
    Gas-Cap Expansion Drive 9
    Water Drive 9
    Gas Reservoir Drives 12
  Effects of Reservoir Drive Mechanisms 12
    Dissolved-Gas Reservoirs 12
    Gas-Cap Expansion Reservoirs 14
    Water Drive Reservoirs 15
    Combination Drive Mechanisms, Hawkins Field Example 17
  Multiple Reservoirs 18
3. WELL COMPLETIONS 19
  Open-Hole Completions 19
  Perforated Casing Completions 20
    Perforated Liner Completions 21
    Tubingless Completions 21
  Completion Interval Selection 22
    Dissolved-Gas Drive Reservoirs 22
    Water Drive Reservoirs 22
    Gas-Cap Expansion Drive Reservoirs 23
    Unknown Drive Mechanism Reservoirs 23
  Types of Completion Arrangements 24
    Single Completions 24
    Dual Completions 25
    Triple Completions 26
    Alternate Completions 27
    Tubingless Completions 28

4. EVALUATING WELL PERFORMANCE 31
Periodic Well Tests 31
Productivity Index Method 32
Difficulties with PI Method 35
Vogel's Equation 36
Selecting the Correct Evaluation Method 39
Dissolved-Gas Drive Reservoirs 39
Water Drive Reservoirs 39
Gas-Cap Expansion Drive Reservoirs 39
General 40
5. PRIMARY CEMENTING 41
Types of Cement 41
Additives 42
Accelerators 42
Retarders 42
Lightweight Additives 44
Heavyweight Additives 46
Lost Circulation Control 46
Filtration Control 46
Salt Cements 46
Silica Flour 47
Factors That Affect Slurry Design 47
Pressure, Temperature, and Pumping Time 47
Cement Viscosity 47
Strength of Cement 48
Slurry Density 48
Resistance to Downhole Brines 48
Filtration Control 48
Cement Slurry Volume and Density Calculations 48
How Cement Slurries Are Specified 48
Cement Slurry Yield and Density Calculations 48
Cement Volume Calculations 50
Placement Techniques 51
Displacement and Bonding Problems 51
Solutions to Displacement and Bonding Problems 52
6. PERFORATING 53
Jet Perforating Charges 53
Principle of Operation 53
Effect of Charge Size on Penetration 54
Penetration Tests 54
Types of Perforating Guns 56
Hollow Steel Carrier Guns 57
Fully Expendable Guns 57
Semiexpendable Guns 57
Gun Selection 58
Factors Affecting Jet Perforating Performance 58
Size and Design of Charges 58
Standoff Problem 58
Hole Conditions 59
Penetration Depth and Shot Density 59
Shot Orientation 60
Casing Damage Potential 61

Optimum Conditions for Perforating . . . 61
Perforating Program Selection . . . 62
Perforating High-Permeability Formations . . . 62
Perforating Low-Permeability Formations . . . 63
7. SQUEEZE CEMENTING . . . 65
High-Pressure Squeeze . . . 65
Development of Cement Squeeze Concepts . . . 65
Low-Pressure Squeeze . . . 66
Placement Techniques . . . 68
Bradenhead Method . . . 68
Packer Method . . . 68
Designing a Cement Squeeze Job . . . 69
Cement Slurry Design . . . 69
Fracture Gradient Considerations . . . 69
Block Squeeze . . . 73
Circulation Squeeze . . . 74
8. PACKER AND TUBING FORCES . . . 75
Types of Packers . . . 75
Retrievable Packers . . . 75
Permanent Packers . . . 77
Forces Acting on Packer-Tubing Systems . . . 77
Forces Acting on Packer Body . . . 77
Forces Acting on Tubing . . . 79
Anchored vs. Unanchored Tubing . . . 85
Packer-Tubing Calculations . . . 87
Retrievable Weight-Set Packer Calculations . . . 87
Permanent Packer Calculations . . . 89
Temperature Profiles in Tubing . . . 89
Prevention of Buckling . . . 90
Helical . . . 90
Mechanical . . . 91
9. PROBLEM WELL ANALYSIS . . . 93
What Problem Well Analysis Is . . . 93
Well Analysis Tools . . . 93
Well Performance Curves . . . 93
Well Status Maps . . . 97
Well Histories . . . 98
Wellbore Sketches . . . 98
Bottomhole Pressure Data . . . 99
Fluid Analyses . . . 99
Fluid Levels . . . 99
Other Well Analysis Tools . . . 100
Problem Well Analysis Examples . . . 100
Declining Oil Production from Oilwell . . . 100
Declining Gas Production from Gas Well . . . 102
10. WORKOVER METHODS . . . 105
Conventional Workovers . . . 105
Well-Killing Procedures . . . 105
After Well Killing . . . 106
Unconventional Workovers . . . 106
Concentric-Tubing Workovers . . . 107
Coiled-Tubing Workovers . . . 107

Wireline Workovers . . . 109
Through-Flowline Workovers . . . 109
Example Problems in Workover Methods . . . 110
Sand Fill Problem . . . 110
Squeezing Off Perforations and Recompleting . . . 111
Choosing the Optimum Workover Method . . . 114
Snubbing Units . . . 114
Principle of Operation . . . 115
Application . . . 116
11. WORKOVER PLANNING . . . 117
Workover Planning Considerations . . . 117
Evaluation of Future Depletion of Wells . . . 117
Evaluation of Competitive Position . . . 118
Workovers to Reduce Water Production . . . 120
Workovers to Reduce Gas Production . . . 121
Workovers to Eliminate Extraneous Water Production . . . 122
Special Problems with Workovers . . . 122
Selection of Workover Method . . . 122
Writing Workover Procedures . . . 122
Basic Well Data . . . 122
Wellbore Sketch . . . 123
Pertinent Well History . . . 123
Reason for Workover . . . 123
Current Well Tests . . . 123
Procedure to Be Followed . . . 123
Well-Testing Procedure after Workover . . . 123
12. BEAM PUMPING IN ARTIFICIAL LIFT . . . 125
The Beam Pumping System . . . 125
Surface Equipment . . . 125
Downhole Equipment . . . 127
Subsurface Pumps . . . 128
Operation . . . 128
Pump Displacement Calculations . . . 128
Rod and Tubing Effects . . . 129
Tubing Stretch . . . 129
Rod Stretch . . . 130
Plunger Overtravel . . . 130
Pumping Unit Load Calculations . . . 131
Pump Selection and Rod String Design . . . 131
Mills Method . . . 136
API RP 11L Method . . . 138
Example Problem . . . 147
Mills Solution . . . 147
API RP 11L Solution . . . 149
Comparison of Results . . . 152
Glossary . . . 153

PREFACE

Petroleum Production Operations was written to fill the need for a particular level of description and explanation of the activities involved in getting petroleum from beneath the ground to the surface. The information that it presents will be readily understandable to those who have been actively involved in the processes of petroleum production on a field supervisory level or to those who are taking or have taken courses in general petroleum technology. The book is not aimed toward the novice; neither is it intended for the petroleum engineer. It is meant to be a basic overview text for those with backgrounds in between.

The first two chapters of the book concern basic facts necessary for understanding natural forces in petroleum reservoirs and reservoir drive mechanisms. Chapter 3 covers well completion, the first step in production; the next chapter looks at evaluating well performance. Chapters 5, 6, and 7 pertain to the processes of primary cementing, perforating, and squeeze cementing. Chapter 8 discusses packer and tubing forces; chapter 9, problem well analysis. Workover methods and workover planning provide the focus for chapters 10 and 11; and beam pumping, as the most popular form of artificial lift, is the subject of the final chapter.

It is to be hoped that neither errors nor outdated content are present in *Petroleum Production Operations*, but if either is detected by the reader, PETEX shall be most grateful for being apprised of it.

Jodie Leecraft
Editor

ACKNOWLEDGMENTS

It is with deep gratitude that we acknowledge the following persons for their careful and expert review of the manuscript of *Petroleum Production Operations:*

Larry M. Bell, Larry V. Macichek, and engineering staff of ARCO Exploration and Technology Company, Dallas, Texas
Myron H. Dorfman, Professor of Petroleum Engineering, The University of Texas at Austin
Walter A. Frnka, Texaco USA, Midland, Texas
J. W. Hargis, Texas Offshore, Mobil Producing Texas and New Mexico, Inc., The Woodlands, Texas

In addition, we thank the following associations and companies for their generosity in allowing us to reprint material or in providing photographs:

American Petroleum Institute, Production Division, Dallas, Texas
Baker Packers Training Center, Houston, Texas
Bethlehem Steel Corporation, Supply Division, Tulsa, Oklahoma
Halliburton Services, Duncan, Oklahoma
Oil and Gas Consultants International, Inc., Tulsa, Oklahoma
PennWell Publishing Company, Tulsa, Oklahoma
Schlumberger Well Services, Houston, Texas
Society of Petroleum Engineers of AIME, Richardson, Texas

I FUNDAMENTALS OF PETROLEUM RESERVOIRS

Fundamental understanding of petroleum reservoirs concerns reservoir rock permeability, reservoir fluid, and reservoir fluid flow concepts. Deeper understanding is generally considered the province of reservoir engineers. However, a basic knowledge of reservoir behavior is needed to make intelligent well completions. Will the well flow to depletion, or will it require artificial lift? Will it produce large volumes of water or gas? Such are the questions that arise in completing a well, and they affect the type of completion made.

A knowledge of reservoir behavior is also needed to evaluate well tests and predict future well performance. Normal well performance needs to be understood so that abnormal or anomalous behavior can be recognized. This basis of problem well analysis is used to determine the need for well workovers.

Reservoir Rock Permeability

Most commercial oil and gas accumulations occur in the pore space of sandstone, limestone, or dolomite rocks. Reservoir fluids, in order to be produced, have to flow through the pore space of the rock. As a filter made by packing a piece of pipe with fine sand offers resistance to the flow of water, so does rock when reservoir fluids (oil, water, and gas) flow through it. The fluids encounter resistance as they flow through the tortuous paths of pore spaces.

Absolute Permeability

The resistance to flow through reservoir rocks is also analogous to resistance in electric flow. In reservoir engineering, the term *permeability* denotes the ease with which fluids flow through a reservoir rock. Permeability is analogous to the electrical flow term *conductance.* In routine core analysis tests, permeability is usually measured by making air flow through a dried sample of rock. Permeability data on routine core analysis tests are usually recorded in millidarcys. One darcy is equal to 1,000 millidarcys.

The abbreviation for millidarcy is *md.* The formal definition of 1 darcy is that permeability which will result in the flow of 1 cubic centimetre per second of a fluid of 1 centipoise viscosity a distance of 1 centimetre through an area of 1 square centimetre under a differential pressure of 1 atmosphere.

Theoretically, absolute permeability is a property of the rock and not of the fluid that flows through it, *provided* that the rock is saturated 100% by the flowing fluid. Since routine core analysis is performed by making air flow through a cleaned and dried sample of rock, this condition is met during routine core analysis. As a practical matter, the air permeability values have to be reduced in evaluating the flow of oil through reservoir rock by using routine core analysis permeability data. A useful rule of thumb is to reduce the air permeability values by 40% to 50%. For example, if routine core analysis shows the air permeability of a rock to be 150 md, a value of 75 to 90 md would be used in evaluating the flow of oil.

Relative Permeability

Absolute permeability is a measure of the ability of a rock to transmit fluid. Unfortunately, complicating factors make it necessary to introduce another term, *relative permeability*. As mentioned above, routine core analysis tests are performed by making air flow through a dried specimen of the rock. While this is a good basis for comparing different rocks in core analysis, the fluids that flow through the reservoir rocks in production operations are oil, water, and gas. A number of complicating factors affect the flow of these fluids through the rock.

The first thing to consider is connate water. Oil and gas deposits are normally laid down in the presence of salt water, and quite often the rock is water-wet. If the rock is a water-wet sandstone, connate water will coat the sand grains with a thin film of water, and the water will also be "wicked" by capillary action back into the small pore spaces of the rock that do not contribute to flow. Connate water cannot be produced, regardless of how much pressure differential is placed across the rock. A typical sandstone rock might initially contain 25% connate water and 75% oil in its pore space. Only the oil will flow to the wellbore when a well is drilled to the reservoir to create a pressure sink. Theoretically absolute permeability should be determined by flowing oil through a rock specimen 100% saturated with oil. As a practical matter, the absolute permeability obtained by flowing oil in the presence of connate or irreducible water saturation gives approximately the same value. Oil flowing in the presence of connate water is normally assumed to have a permeability equal to the absolute permeability.

Another complicating factor is two-phase flow. Although it is desirable to produce only oil from a reservoir, oil and gas are often produced simultaneously. The gas interferes with the flow of oil; therefore, the permeability to oil is reduced. The more free gas present, the more the permeability to oil is reduced. In order to evaluate the effect of gas on oil permeability, *relative permeability curves* (fig. 1.1) must be determined experimentally.

Notice that k_o/k, the ratio of the oil permeability to the absolute permeability and k_g/k, the ratio of the gas permeability to the absolute permeability, are plotted against liquid saturation. At 100% liquid saturation, the ratio of k_o/k is 1 by definition. This condition exists when only oil is flowing and no gas is present to interfere with the flow of oil. If the absolute permeability of the rock is obtained by making oil flow through it in the presence of only connate water, then k_o/k will equal k, and k_o/k will be equal to 1, the value at 100% liquid saturation. If a free gas saturation of 10% is introduced, the liquid saturation will be reduced to 90%. Figure 1.1 shows that the k_o/k value will be decreased to 0.7 at a value of 90% liquid saturation. For a rock with an absolute oil permeability of 1,000 md, the permeability to oil will be reduced to 700 md at a gas saturation of 10%.

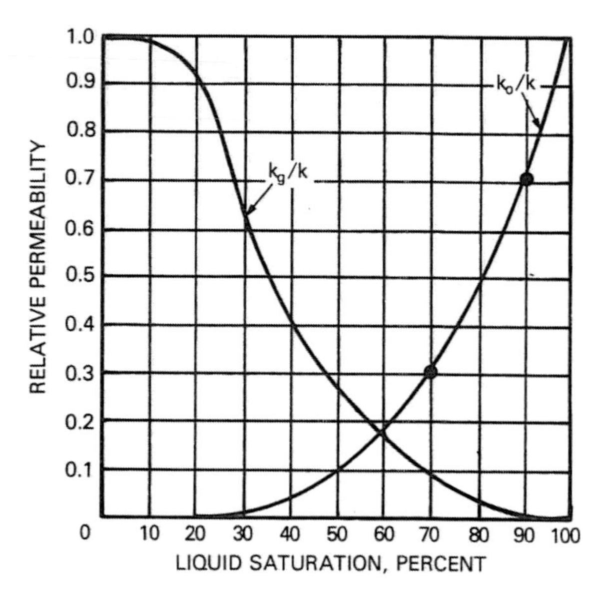

Figure 1.1. Relative permeability curves

If the gas saturation is increased to 30%, the value of k_o/k will be reduced to 0.3. The value of k_o at 30% gas saturation for this hypothetical rock is then equal to 300 md ($k_o = 0.3 \times 1{,}000 = 300$ md). How is the gas saturation increased? In an oil reservoir an increase in gas saturation is basically due to a decrease in pressure, which occurs with production.

The above curves show how the permeability of each phase is dependent on its saturation in the pore space. Note that the k_o/k curve starts at the lower end at about 20%. This is the connate water saturation for this sample. The rest of the liquid saturation, 80%, is made up of oil. As oil is produced, it is replaced by free gas.

The above curves are called *gas-oil relative permeability curves.* There are also water-oil relative permeability curves that cover the simultaneous flow of oil and water in a reservoir rock; they are interpreted in a similar manner.

Reservoir Fluids

Solubility of Gas in Crude Oil

The important characteristics of crude oil are dependent on its ability to dissolve gas as the pressure is increased. Whenever oil is present in a reservoir, it is accompanied by solution gas. Although the amount of gas in solution may be as little as 14 cubic feet per barrel, more typically it is in the range of 500 to 600 cubic feet per barrel. The amount of gas in solution affects the amount of space the oil occupies in the reservoir and also affects the oil viscosity.

The amount of gas that will go into solution in oil increases as the pressure increases (fig. 1.2). At a pressure of 2,500 psia, 567 cubic feet of gas is dissolved in each barrel of oil. As the pressure is decreased to 500 psia, the amount of gas in solution is reduced to only 200 cubic feet per barrel.

According to Henry's law, the amount of gas going into solution in a liquid is directly proportional to the pressure exerted by the gas. Unfortunately, this law does not apply to crude oils, and solution gas-oil ratio curves, such as the example above, must be determined experimentally. A sample of oil at bottomhole pressure and temperature is obtained, and the solubility curve is constructed by measuring the amount of gas liberated as the pressure is reduced.

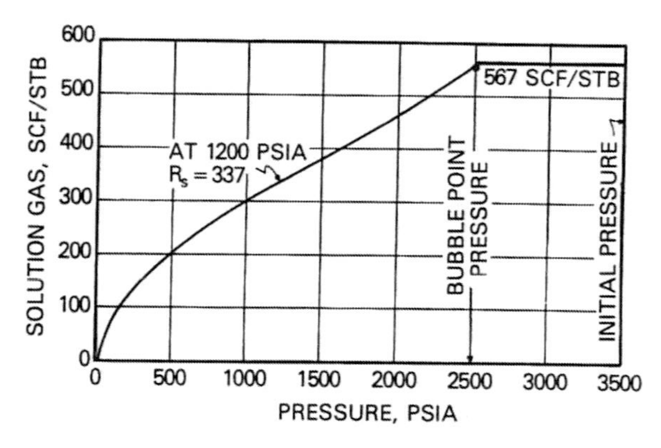

Figure 1.2. Solution gas-oil ratio, Big Sandy Field, at a reservoir temperature of 160° F. From B. C. Craft and M. F. Hawkins, *Applied Petroleum Reservoir Engineering* (Englewood Cliffs, N. J.: Prentice-Hall, 1959).

The curve (fig. 1.2) has a flat portion at the higher pressures. The crude is called *undersaturated.* The undersaturated state indicates that there is a deficiency of gas present. If gas had been sufficient, it would have continued to go into solution, and the curve would not have leveled off at the higher pressures. Reservoir oil may also have been fully saturated; the oil may have absorbed all of the gas it was capable of holding in solution, regardless of pressure.

Note the initial pressure end of the curve. As the pressure is reduced to the *bubble point* pressure, no change occurs. The bubble point pressure for an undersaturated crude is the pressure at which the first bubble of free gas appears. In a sample of crude under initial pressure in a transparent cylinder, the bubble point can be observed by slowly lowering the pressure until the first bubble of gas becomes visible.

Note further that as the pressure is lowered below the bubble point, the amount of gas remaining in solution decreases. Where does the gas that was liberated go? In the reservoir, the gas that is liberated occupies the space that is vacated by the produced oil. It is called *free gas,* and it flows to the wellbore along with the produced oil. It is this free gas that interferes with the flow of oil and causes a decrease in oil permeability.

Remember that as the pressure is decreased, the amount of gas remaining in solution is decreased. This factor is important in understanding the formation volume factor and the viscosity of crude oil.

Formation Volume Factor

Oil and gas are produced simultaneously from an undersaturated oil reservoir. As the oil and gas reach the surface, they are separated in a separator; the gas goes to sales and the oil goes to the stock tank. Of course, if the lease is automated, the oil goes directly to the pipeline. It should be obvious that the volume of the oil in the stock tank is going to be less than the volume that the oil occupied in the reservoir when it still had gas in solution. The factor that is used to convert stock tank barrels to reservoir barrels is called the *formation volume factor (FVF)*, or *reservoir volume factor (RVF)*. The FVF is defined as the ratio between the space occupied by a barrel of oil containing solution gas at reservoir conditions and a barrel of dead oil at surface conditions, usually measured at 14.7 psia and 60° F. Formation volume factor curves are determined experimentally (fig. 1.3). The curves in figures 1.2 and 1.3 were determined from the same crude oil sample.

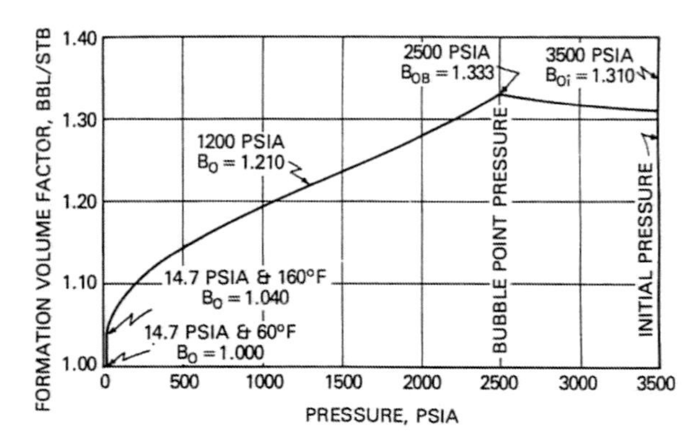

Figure 1.3. Formation volume factor of the Big Sandy Field reservoir oil at a reservoir temperature of 160° F. From B. C. Craft and M. F. Hawkins, *Applied Petroleum Reservoir Engineering* (Englewood Cliffs, N. J.: Prentice-Hall, 1959).

The curve in figure 1.3 is very similar to the solution-gas curve because the shape of both curves is dependent upon the amount of gas remaining in solution in the reservoir oil at a given pressure. A difference, however, is that while the solution-gas curve is flat above the bubble point, the FVF curve decreases as the pressure increases above the bubble point. The gas in solution is compressed due to the increased pressure, and the oil with dissolved gas occupies less space.

The curve at a pressure of 500 psia reads a formation volume factor of 1.14 reservoir bbl/stock tank bbl. Remember that each barrel of oil contains 200 cubic feet of dissolved gas at a pressure of 500 psia. Assume the pressure is 1,000 psia; the FVF has increased to 1.19. Check the solution gas-oil ratio curve at 1,000 psia; there are now 300 cubic feet of gas in each barrel of oil. The addition of the extra 100 cubic feet of gas dissolved in each barrel has increased the FVF from 1.14 to 1.19.

Another way of looking at the FVF is to assume a crude with an FVF of 1.25. An FVF of 1.25 would indicate that 1.25 barrels of oil in the reservoir would shrink to 1.0 barrel in the stock tank. This reduction can be expressed as a *shrinkage factor*. The shrinkage factor is $1.0/1.25 = 0.8$. The oil shrinks so that the oil in the stock tank occupies only 80% as much space as it did in the reservoir. Normally an oil with an FVF of 1.4 would be considered a high-shrinkage crude, while an oil with an FVF of 1.2 would be considered a low-shrinkage crude.

Oil Viscosity

Dissolved gas has another important effect on the viscosity of crude oil. *Viscosity* is a measure of the tendency of a liquid to resist flow. The viscosity of an oil in the stock tank is much higher than the same oil was in the reservoir. Both the temperature and the pressure are lower at the surface than they are in the reservoir. Lower temperature causes higher viscosity. But lower pressure causes lower viscosity. The two factors act in opposition to one another.

However, the final and by far the most important effect on oil viscosity is the amount of dissolved gas. The difference in oil viscosity between stock tank and reservoir conditions is mainly due to the amount of dissolved gas present in the reservoir oil (fig. 1.4).

Gas solubility and FVF curves must be remembered in looking at the oil viscosity plot, since they are influenced by the amount of gas in solution. The FVF and gas solubility decrease as the pressure is lowered below the bubble point, while the oil viscosity increases–all because of the reduced volume of gas in solution in the oil. Since the FVF, gas solubility, and reservoir oil viscosity curves change with pressure, they are said to be *pressure dependent.* Oil viscosity is measured in centipoises (cp) for use in calculations where oilfield units are used.

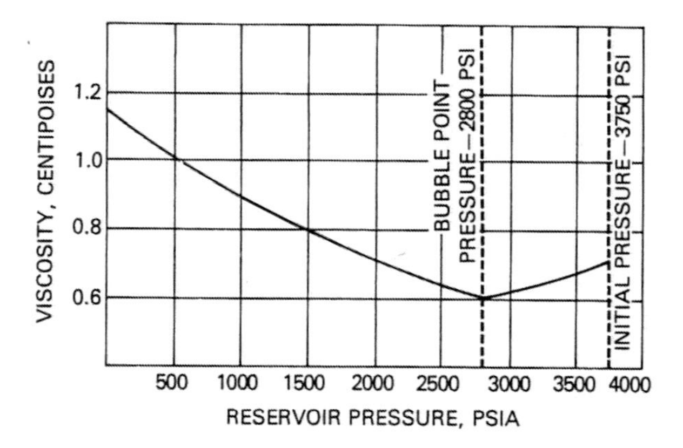

Figure 1.4. Change of reservoir oil viscosity with change in pressure

Fluid Flow in the Reservoir

Oil accumulations are found in reservoir rock under pressure. Normal reservoir pressure is approximately 0.47 psi per foot of depth. A normally pressured reservoir at a depth of 10,000 feet should have an initial pressure of about 4,700 psi. After a well is drilled into an oil reservoir, equipped with casing and tubing, perforated, and placed on production, a pressure sink is created at the wellbore due to fluid withdrawals at this point. The term *pressure sink* means a condition in which the pressure at the wellbore is less than the reservoir pressure. Flow of oil from the reservoir to the wellbore occurs because of this pressure differential. For the same reason, there is flow in a pipeline. When the pressure is increased at one end, the oil flows out the other end where pressure is lower. The difference in pressure furnishes the driving force for the flow.

Radial Flow

Flow through a reservoir is different from that through a pipeline. In addition to the effects of permeability, oil flow through a reservoir is complicated by the radial-flow phenomenon. In a pipeline, flow is linear through a constant cross-sectional area. In the case of an oilwell, oil flows into the wellbore from all directions–hence the term *radial flow*. The cross-sectional area available to flow constantly decreases as the oil approaches the well, assuming that the reservoir is uniform and of constant height. The flow area can be visualized as a series of concentric cylinders of constant height and constantly decreasing diameter as the well is approached. Near the wellbore, the cross-sectional area available for flow decreases, since the diameters of the circles decrease and the height is constant. This problem becomes acute in the near-wellbore area (fig. 1.5).

Most of the pressure drop occurs near the well; the outer boundary of the well's drainage radius is relatively unaffected. The drainage area is the area affected by the pressure sink created by the well.

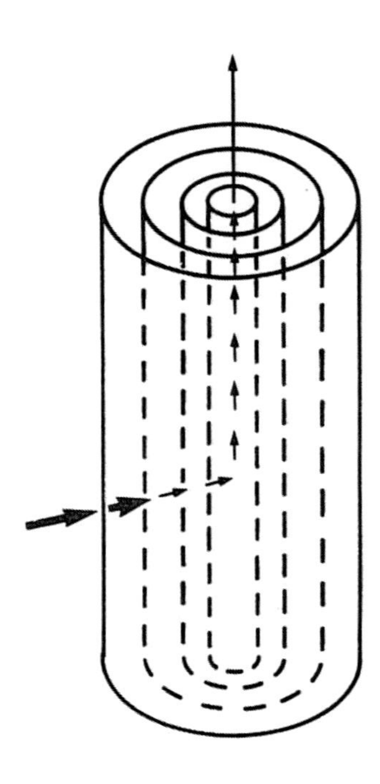

Figure 1.5. Radial flow in a reservoir

Darcy's Law

Radial flow can be evaluated if certain basic rock and fluid data are available. The tool used to make this evaluation is called Darcy's law. The following form of Darcy's law is valid for steady-state radial flow of incompressible fluids, using conventional oilfield units.

$$q = \frac{7.08\, kh\,(p_e - p_w)}{B\mu\, ln\, r_e/r_w}$$

where

q = stock tank barrels/day
k = permeability, darcys
h = reservoir thickness, feet
B = formation volume factor, vol/vol
μ = oil viscosity, centipoises
ln = natural logarithm
p_e = pressure at drainage boundary, psi
p_w = wellbore flowing pressure, psi
r_e = radius of drainage, feet
r_w = wellbore radius, feet.

Note that the equation calls for permeability in darcys. Permeability is normally reported in millidarcys, so they must be converted to darcys before the Darcy equation is solved.

This equation contains the various factors that influence a well's producing rate. Note that the flow rate varies directly with the permeability, the reservoir height, and the term $p_e - p_w$; $p_e - p_w$ is called the *pressure drawdown,* or just *drawdown.* Doubling the permeability will double the flow rate if all other conditions remain the same. The flow rate can also be doubled by doubling the formation height or the drawdown, if the other factors remain the same. Obviously, the reservoir height is fixed, but permeability can be increased by acidizing or fracturing the well. The wellbore pressure can be lowered and the drawdown increased by increasing the choke size if the well flows, or artificial lift can be added.

Note also that the flow varies inversely with the FVF and reservoir oil viscosity. These two values cannot normally be changed, although the viscosity can be lowered in some wells by thermal processes such as steam injection.

A review of the Darcy equation produces a qualitative feel for the factors that affect an oilwell's producing rate. Oil flow rate can also be solved for if the required basic data is at hand.

To illustrate the use of Darcy's equation, assume that a well has just been drilled and it is desirable to evaluate how much it will produce. The well logs show that the pay thickness, h, is 25 feet. Routine core analysis data indicate that the air permeability in the pay zone is 300 md. From oil analysis data an estimate of the oil viscosity, μ, is calculated to be 0.63 cp and the FVF to be 1.24. A drill stem test shows that the static reservoir pressure, p_e, is 2,100 psi, and the flowing bottomhole pressure, p_w, is estimated to be 1,600 psi. Since the well is completed in a field with 80-acre spacing, a drainage radius of 1,050 feet can be estimated. (One acre = 43,560 square feet. $r^2 = 80 \times 43{,}560/\pi$; $r = 1{,}050$ feet. This calculation assumes a circular drainage pattern.) The caliper log shows that the wellbore diameter is 9 inches, or 0.375 feet.

The value of air permeability obtained on routine core analysis must be reduced by 40% to 50%, since oil flow is being evaluated. The air permeability will be reduced by 50% in the present calculations. (300 md × 0.5 = 150 md = 0.15 darcys.)

Using the above data, the flow rate of the well is estimated as follows:

$$q = \frac{7.08 \times 0.150 \times 25 \times (2{,}100 - 1{,}600)}{1.24 \times 0.63\, ln\, 1{,}050/0.375}$$

$$q = 2{,}141 \text{ stock tank barrels/day.}$$

The well should flow about 2,000 to 2,100 barrels per day based on the above calculation. If the above numbers are punched into a calculator, the solution will be 2,141 barrels/day, as indicated above. The data used are not accurate enough to give an answer to four significant figures.

II Reservoir Drive Mechanisms and Their Effects

A basic knowledge of reservoir drive mechanisms is needed to make intelligent decisions regarding well completions. A knowledge of reservoir behavior is also needed in many other aspects of production operations such as artificial lift, problem well analysis, workover planning, and well performance evaluation.

Sources of Reservoir Energy

Only three natural sources of energy are available to move reservoir fluids (oil, water, and gas) to the wellbore – fluid expansion, rock expansion, and gravity. Reservoir rocks and fluids are under pressure because they are normally encountered at depths where they lie under several thousand feet of sediments. When a well penetrates a new reservoir, pressure will be encountered, since the oil, water, gas, and rock are in a compressed state. Water and rock are normally considered to be incompressible, but actually both are slightly compressible. Therefore, the amount of reservoir rock and water directly connected to the reservoir will determine how much they affect the reservoir drive mechanism. The fact will be more apparent as the three oil reservoir drive mechanisms are examined in detail.

Types of Reservoir Drives

Three reservoir drive mechanisms are basic to oil reservoirs – dissolved-gas, or volumetric, drive; water drive; and gas-cap expansion drive. Gas reservoirs have either volumetric or water drive. Since the characteristics of each drive mechanism are different, they will affect decisions on the location of wells and the selection of completion intervals.

Dissolved-Gas Drive

Dissolved-gas drive is sometimes referred to as volumetric drive, internal-gas drive, or depletion drive. The term *dissolved-gas drive* is the most descriptive of the process involved. The dissolved-gas drive is sometimes explained by a soda pop bottle analogy. If a bottle of carbonated soft drink is shaken with a thumb over the top, pressure will be developed from the carbon dioxide dissolved in the liquid, and the liquid can be sprayed from the bottle. As soon as the supply of carbon dioxide is exhausted, the process stops, and only dead liquid is left in the bottle. The depletion of a dissolved-gas drive oil reservoir is similar although a little more complex, since the expansion of the reservoir rock and connate water plus the relative permeability effects complicate the process.

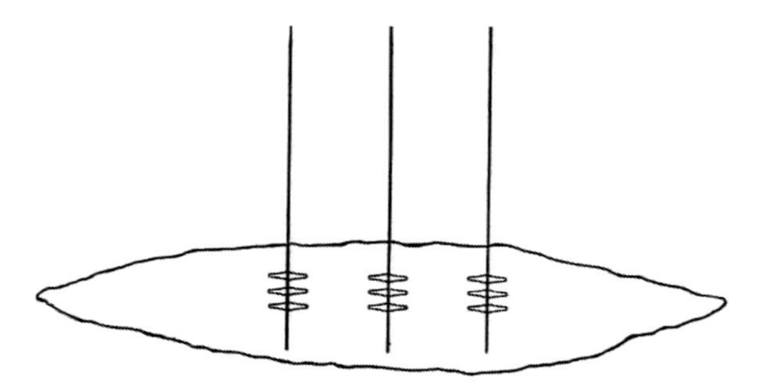

Figure 2.1. Typical dissolved-gas drive reservoir

A dissolved-gas drive oil reservoir is one in which the oil-filled rock is not in contact with an aquifer (fig. 2.1). The oil-filled rock can be isolated by shale barriers or permeability-porosity pinchouts.

A dissolved-gas drive oil reservoir can be thought of as a closed tank filled with porous rock (fig. 2.2). The pore space of the rock is initially filled with connate water and oil containing dissolved gas. The oil with the dissolved gas is normally an undersaturated crude. The entire reservoir is under pressure when it is penetrated by a well. A pressure sink is created at the wellbore when the well is perforated, and oil will flow from the reservoir rock into the well. Initially, the flow of oil will be sustained by the expansion of the liquid (oil and water) and the rock; it is the source of energy until the bubble point pressure is reached, but it has a negligible effect on production below the bubble point. Gas expansion is mainly responsible for production below the bubble point because the gas phase has higher compressibility. Gravity has a negligible effect on the recovery of oil from the typical dissolved-gas drive reservoir.

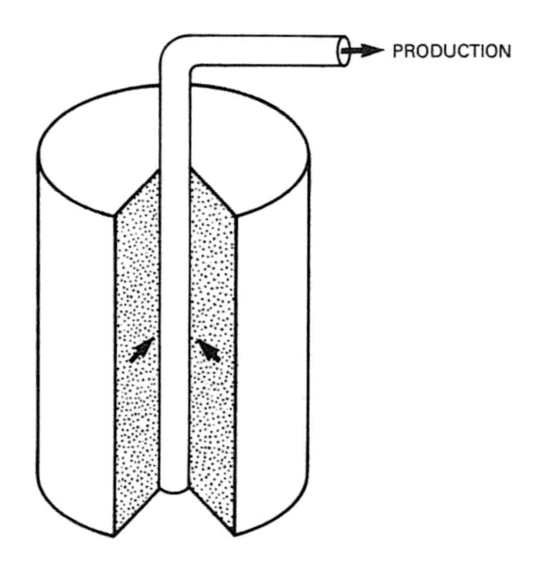

Figure 2.2. Closed-tank analogy, dissolved-gas drive reservoir

Oil recoveries from dissolved-gas drive reservoirs usually range from 15% to 25% of the original oil in place. Assuming a median recovery of 20%, 2% to 3% would typically be due to liquid and rock expansion and the remaining 17% to gas expansion.

The low recovery from a dissolved-gas drive reservoir is due to the inefficient use of the expansive energy of the gas. As the bubble point pressure is reached, gas comes out of solution and exists as unconnected bubbles in the reservoir. At this stage, oil is displaced to the wellbore in an efficient pistonlike manner by the expansion of the gas. As the reservoir pressure decreases, the gas bubbles become connected and a continuous free-gas saturation flows to the wellbore. Since the gas flows directly to the wellbore, much of its expansive energy is lost. The simultaneous flow of gas and oil decreases the relative permeability to oil (fig. 1.1). The process is accelerated as the pressure drops so that the flow of gas increases and the flow of oil decreases. This effect is illustrated graphically in figure 2.3, a typical performance curve for a dissolved-gas drive reservoir.

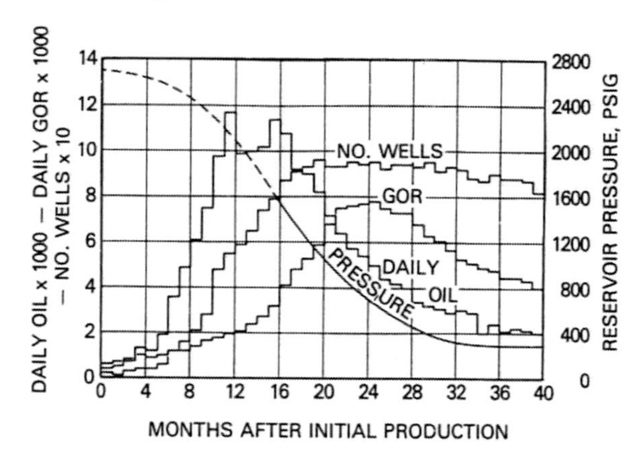

Figure 2.3. Typical performance curve for a dissolved-gas drive reservoir. From B. C. Craft and M. F. Hawkins, *Applied Petroleum Reservoir Engineering* (Englewood Cliffs, N. J.: Prentice-Hall, 1959).

Once the free gas starts flowing to the wellbore, the drop in reservoir pressure increases. More gas comes out of solution in the oil remaining in the reservoir, and the producing gas-oil ratio increases. The decline of the production rate is accelerated, and the reservoir is depleted as in the soda pop analogy.

Gas-Cap Expansion Drive

The true gas-cap expansion drive is not common, and several special conditions must exist for this type of drive to develop.

1. The reservoir must be steeply dipping.
2. The reservoir rock must have relatively high vertical permeability.
3. The gas cap must be very large in relation to the oil column.
4. The reservoir must be produced at a rate that permits gravitational segregation of the oil and gas.

A gas-cap expansion drive reservoir is shown schematically in figure 2.4. Note that all producing wells are completed in the oil columns below the gas cap.

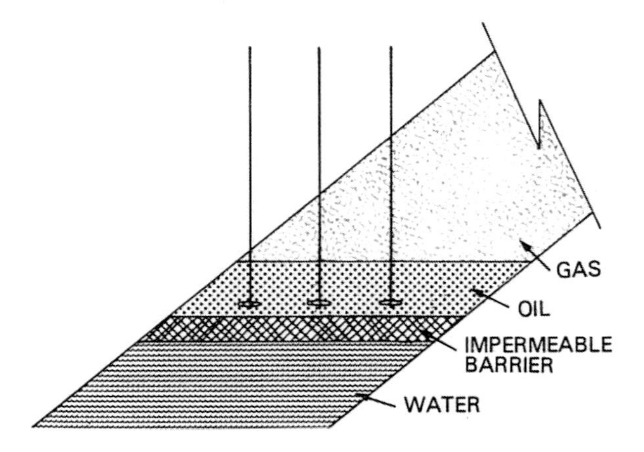

Figure 2.4. Typical gas-cap expansion reservoir

The production of oil reduces the pressure in the oil column. The gas cap expands to fill the void left by the production of oil, and the expansion of the gas cap helps to maintain the pressure in the oil column. Since the pressure declines at a slow rate, the gas-oil ratio of the producing wells remains relatively constant. Eventually, as oil is displaced down to the first producing well, gas-cap gas breaks through to the wellbore, and the well will quickly start producing mostly gas. A typical performance curve for a gas-cap expansion drive field is shown in figure 2.5.

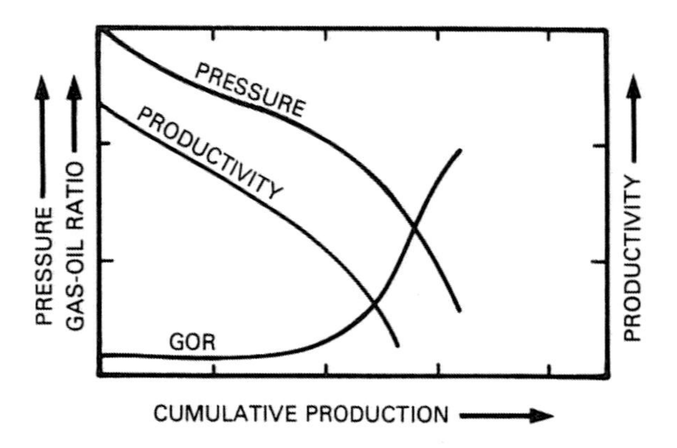

Figure 2.5. Typical performance curve for a gas-cap expansion drive field. From Kermit E. Brown, *The Technology of Artificial Lift Methods*, vol. 1 (Tulsa: PennWell Publishing Company, 1977).

Recovery from gas-cap expansion drive fields can be very high; recoveries as high as 85% have been recorded. But in order to achieve a high recovery rate in a gas-cap expansion drive field, it is usually necessary to reinject all of the produced gas as well as some extraneous gas back into the gas cap to help maintain gas-cap pressure.

Water Drive

The energy to sustain production in a water drive field comes from the expansion of water. In the discussion of dissolved-gas drive reservoirs, it was pointed out that only 2% to 3% of the recovery is due to oil, water, and rock expansion. A dissolved-gas drive reservoir is like a closed tank; it is not in contact with a large aquifer. An aquifer is a water-filled porous and permeable rock. In the case of a water drive oil reservoir, the oil-filled reservoir rock is hydraulically connected to a large water-filled reservoir (figs. 2.6 and 2.7).

The coefficient of expansion for water is very low, with 3.0×10^{-6} psi^{-1} being a typical value.

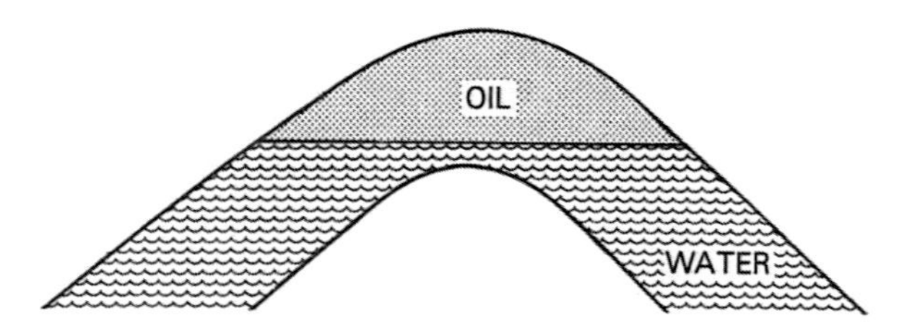

Figure 2.6. Typical water drive reservoir

This means that 1 million barrels of water will expand by 3 barrels if the pressure is reduced 1 psi. It is obvious from a review of these data that a very large aquifer must be in contact with the oil reservoir in order to furnish enough energy to help maintain the pressure in the oil-filled rock. The process is aided by the expansion of the aquifer rock, since a typical rock has a coefficient of expansion of 4×10^{-6} psi^{-1}.

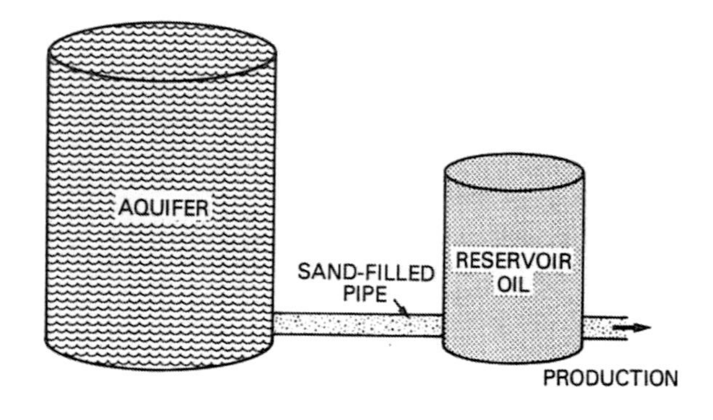

Figure 2.7. Water drive reservoir analogy

The large tank on the left (fig. 2.7) represents the aquifer, and the smaller tank on the right, the oil reservoir. They are connected by a sand-filled pipe. Initially, both tanks are at the same, or original, reservoir pressure. If the pressure in the oil tank is reduced by production, the water and rock in the aquifer tank will expand, and water will flow through the sand-filled pipe into the oil reservoir tank, helping to maintain the pressure in the oil tank. In order for this pressure maintenance to occur, the aquifer must be very large in comparison to the oil reservoir.

The following example illustrates how large an aquifer must be in order for a water drive to be effective. It is not necessary to follow through this example to understand the subsequent material, although it is of interest for presenting an evaluation of rock and fluid expansion.

Assume a circular oil reservoir 5,000 feet in diameter and 30 feet thick (fig. 2.8). The oil reservoir is hydraulically connected to an aquifer 50,000 feet in diameter of the same thickness. Assume the following additional data for the oil reservoir and aquifer:

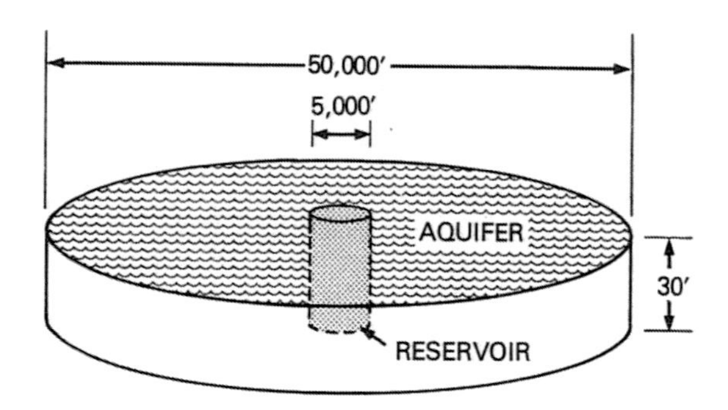

Figure 2.8. Water drive reservoir model

Connate water (S_w) = 25%
Porosity (Φ) = 16%
FVF = 1.25 reservoir bbl/stock tank bbl
Recovery efficiency = 60% (oil reservoir)
Water compressibility = 3×10^{-6} psi^{-1}
Rock compressibility = 4×10^{-6} psi^{-1}
Total compressibility = 7×10^{-6} psi^{-1}

Using the above data, the expansion capability of the aquifer (ΔV) is–

$$\Delta V = 7 \times 10^{-6} \times \pi/4\,(50{,}000^2 - 5{,}000^2) \times 0.16 \times 30$$
$$= 65{,}314 \text{ cu ft/psi}$$
$$= \frac{65{,}314}{5.615} = 11{,}632 \text{ bbl/psi.}$$

In other words, for each 1-psi drop in the aquifer pressure, the rock and water will expand and force 11,632 barrels of water into the reservoir rock.

The hydrocarbon volume can be calculated as follows:

$$V = \pi/4\,(5{,}000^2) \times 30 \times 0.16 \times (1 - 0.25)$$
$$= 70{,}685{,}834 \text{ cu ft}$$
$$= \frac{70{,}685{,}834}{5.615} = 12{,}588{,}750 \text{ reservoir barrels.}$$

Since the recovery factor is 60%, if the oil reservoir is depleted, total oil production will be–

Production = 12,585,750 × 0.6 = 7,553,250 reservoir barrels.

Since the oil shrinks going from the reservoir to the stock tank, the total must be divided by 1.25, the FVF, to get stock tank barrels, as follows:

$$\text{Production} = \frac{7{,}553{,}250}{1.25} = 6{,}042{,}600 \text{ stock tank barrels (STB).}$$

Since the first calculation shows that the aquifer will expand 11,632 bbl for each psi that the pressure is reduced, the total pressure reduction (Δp) required can now be determined.

$$\Delta p = \frac{7{,}553{,}250}{11{,}632} = 649 \text{ psi.}$$

The average pressure would have to be dropped 649 psi to deplete this hypothetical field. Obviously this is a simplification of the actual process, since there will be a pressure gradient across the system. The highest pressure will be at the aquifer boundary and the lowest pressure at the point of oil withdrawal.

A large aquifer is needed. The example illustrates the size of the aquifer needed to sustain a water drive. The aquifer has 1,661,715,081 barrels of water in place. The ratio of aquifer size (A) to oil reservoir size (R) in this example is–

$$A/R = \frac{1{,}661{,}715{,}081}{12{,}585{,}750} = \frac{132}{1}.$$

In the example, the aquifer has to be 132 times larger than the oil reservoir to deplete the oil reservoir with a 649-psi pressure drop. If the pressure drop were limited to 325 psi, then the size of the aquifer would have to be doubled.

The size of the aquifer is just part of the problem. The water must be able to flow into the reservoir at the same rate that oil is being produced. It is obvious that the sand-filled pipe between the two tanks has to be sufficiently large to permit the expansion of the water into the oil tank at the same rate at which it is being taken out (fig. 2.7). A large sand-filled pipe with a high flow capacity is equivalent to a reservoir rock of high permeability. With a low-permeability rock, equivalent to a small-diameter pipe in the tank example, the water cannot expand into the oil tank at the rate necessary to replace the produced oil. In such a circumstance, the oil reservoir behaves like a dissolved-gas drive reservoir. High permeability is the reason for the large numbers of water drive fields in the Frio sands along the Gulf Coast, since typical Frio sands have permeabilities of 500-600 md. Low permeability explains why West Texas limestone and dolomite fields with permeabilities of only 3 or 4 md do not have water drives even if they are connected to large aquifers.

Recoveries from water drive reservoirs are higher than recoveries from dissolved-gas drive reservoirs. Recoveries usually range from 40% to 80%, with 60% recovery being typical for many Gulf Coast fields. Since the pressure is maintained in a water drive field, the gas-oil

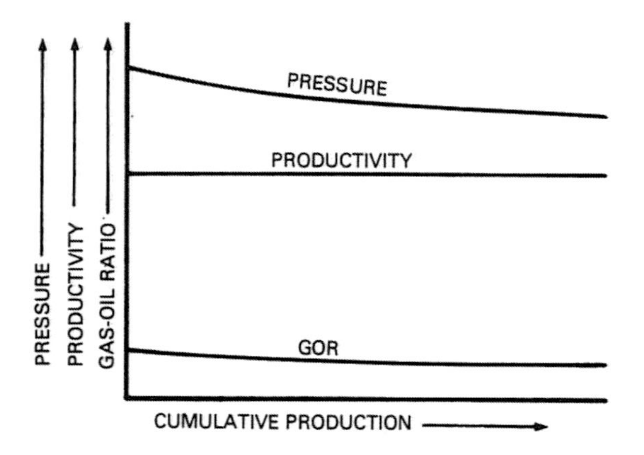

Figure 2.9. Typical performance for a water drive field for a low production rate. From Kermit E. Brown, *The Technology of Artificial Lift Methods,* vol. 1 (Tulsa; PennWell Publishing Company, 1977).

ratio (GOR) remains relatively constant, since the gas is kept in solution (fig. 2.9).

Gas Reservoir Drives

Two types of producing mechanisms for gas reservoirs are volumetric (pressure depletion) and water drive. The volumetric, or depletion, drive reservoir is similar to the dissolved-gas drive oil reservoir except that relative permeability problems usually do not occur. A gas reservoir normally has only one phase. For this reason, volumetric gas reservoirs usually have high recoveries, 75% to 85% being common.

Water drive gas fields are similar to water drive oil reservoirs. The requirements are the same. Recovery from water drive gas fields is about 60%. Unlike oil reservoirs, the recovery from water drive gas fields is considerably less than for volumetric, or pressure depletion, drives. The recovery from water drive gas fields is only about 60% because of a phenomenon called sweep efficiency. The water that flows in from the aquifer to the gas reservoir does not displace the gas uniformly but channels past certain areas due to variations in porosity and permeability. The efficiency with which the water displaces the gas is called *sweep efficiency.* The gas trapped in the unswept areas is essentially at original pressure, resulting in a large volume of gas being trapped. Since the gas trapped in a water drive gas field is at a high pressure, the volume of gas lost is high. The net result is that the recovery from water drive gas fields is usually 60% or less.

Water drive fields can often be produced at rates higher than the water flows in, making the field perform as if it were a volumetric, or pressure depletion, reservoir. This situation is sometimes called "outrunning the water." By producing at high rates, the recovery can be increased from 60% or less to as much as 80% to 85%.

Effects of Reservoir Drive Mechanisms

The type of drive mechanism for a field will determine where wells are located and how they are completed.

Dissolved-Gas Reservoirs

The recovery from a dissolved-gas reservoir is normally not rate-sensitive–that is, the wells can be produced at their highest possible rates without adversely affecting ultimate recovery. Of course, the fields will be depleted more rapidly at the higher rates. GOR performance is normally not rate-sensitive, since the producing GOR is dependent upon the relative permeability relationship of the oil and the gas. Oil and gas relative permeabilities are dependent upon the oil saturation or the stage of reservoir depletion.

The most important factor in a well completion is whether the fields are high-relief (high angle of dip) or low-relief (low angle of dip).

Dissolved-gas reservoir, low-relief. It has been pointed out that in a dissolved-gas reservoir the oil flows from the reservoir to the well primarily because of expansion of the dissolved gas, once the reservoir pressure is below the bubble point pressure. Individual well recoveries in low-relief dissolved-gas reservoirs are based primarily on their flow capacity. The flow capacity of a well, according to Darcy's law, is directly proportional to the permeability (k) and pay thickness (h). Productivity of individual wells in a field then is primarily dependent upon the

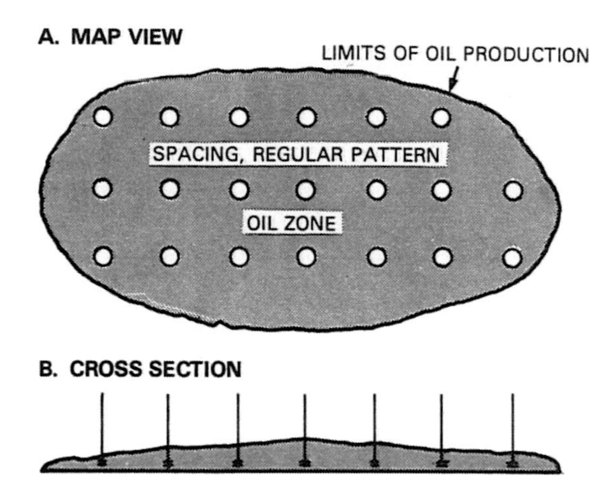

Figure 2.10. Dissolved-gas drive reservoir—low angle of dip. From Norman J. Clark, *Elements of Petroleum Reservoirs* (Dallas: Society of Petroleum Engineers of AIME, 1960).

value of their *kh*. Experience has shown that the ultimate recovery from each well will be proportional to its share of the total field production. Well spacing in this type of field then is normally of a uniform pattern (fig. 2.10).

Generally, the entire pay section is perforated in each well in order to achieve the maximum flow rate possible, since individual well ultimate recoveries will be roughly proportional to their flow rates. Perforating the entire pay section gives the highest flow capacity (*kh*). Some dissolved-gas drive fields are underlain by water, and perforations near the oil-water contact are avoided to minimize the possibility of water production. The recommended procedure for perforating a typical solution-gas drive reservoir well not underlain by water is illustrated by figure 2.11.

A dissolved-gas drive field depletes under primary production, so it will exhibit constantly declining bottomhole pressures and producing rates and constantly increasing GORs. Since primary recoveries are so low for this type of field, some sort of secondary recovery program such as waterflood is usually begun at some point in the life of the field. All of these factors will influence the type of completion that is made.

Dissolved-gas reservoir, high relief. If the high-relief reservoir has good permeability, then the uniform spacing pattern has to be altered.

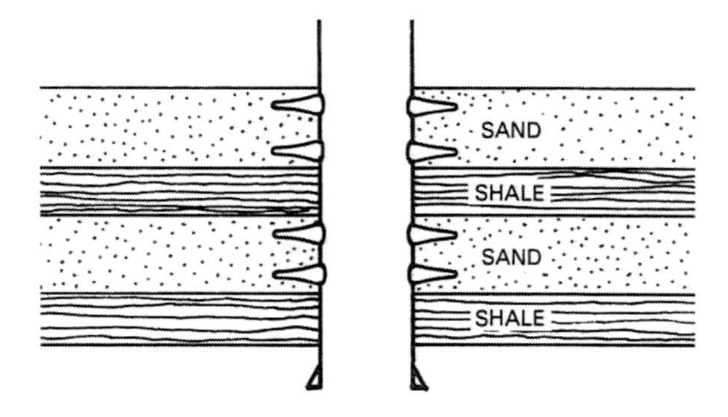

Figure 2.11. Procedure for perforating solution-gas drive reservoir wells

Figure 2.12 depicts a high-relief dissolved-gas drive oil reservoir. If the permeability is high enough to permit gravity segregation of oil and gas, then free gas will migrate to the top of the structure and form what is called a secondary gas cap. Recall that when the bottomhole pressure drops below the bubble point pressure, gas is liberated from the oil.

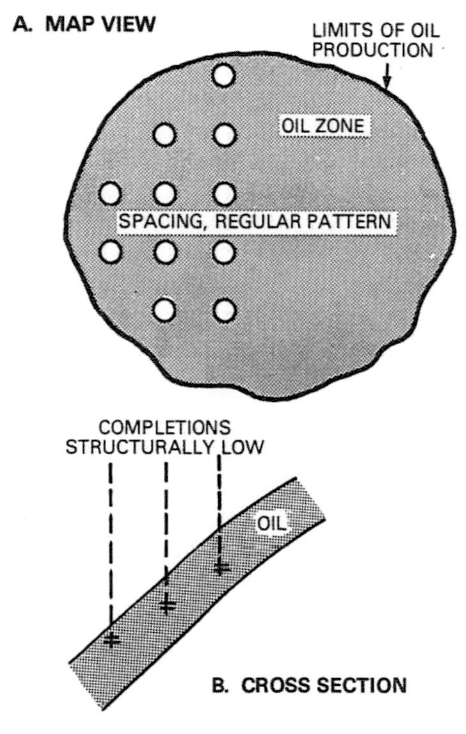

Figure 2.12. Dissolved-gas drive reservoir—high angle of dip. From Norman J. Clark, *Elements of Petroleum Reservoirs* (Dallas: Society of Petroleum Engineers of AIME, 1960).

The formation of a secondary gas cap will usually increase primary recovery slightly, but its main effect will be on the producing gas-oil ratio of any well completed at the top of the structure. Wells completed high on the structure will produce with high GORs as soon as the secondary gas cap forms, and they may go completely to gas as depletion continues. In order to avoid this possibility, wells should not be completed in the updip portion of the reservoir, where a secondary gas cap might develop.

Except for the problems mentioned for upstructure wells, the performance of high-relief dissolved-gas fields normally will be similar to low-relief fields. In other words, the performance of the down-structure wells will be the same as wells completed in a low-relief dissolved-gas drive reservoir.

Gas-Cap Expansion Reservoirs

Ultimate recovery from gas-cap expansion fields can be rate-sensitive. If wells are produced at high rates, gas may channel down into the wellbore and bypass oil in the reservoir, resulting in lower recoveries. The principal type of secondary recovery project for gas-cap fields is reinjecting the produced gas together with extraneous gas in the top of the structure to maintain reservoir pressure. This type of program is especially applicable when the water aquifer starts to encroach from the bottom and displaces oil into the original gas cap. Oil displaced into an original gas cap usually saturates the dry sand grains and becomes nonrecoverable. Such an event can happen, since the reservoir pressure of the gas cap will constantly drop with pressure, sometimes permitting water encroachment from below.

Gas-cap expansion reservoir, low angle of dip. In a low-relief gas-cap drive field (fig. 2.13), the gas cap completely overlies the oil column. As production progresses, the gas cap expands downward, displacing the oil. Since many reservoirs are in contact with an aquifer, some barrier must exist to prevent water influx from making water drive the predominant factor. This barrier may be an asphalt seal at the base of the oil zone. In other cases, the aquifer may be so small that it is ineffective.

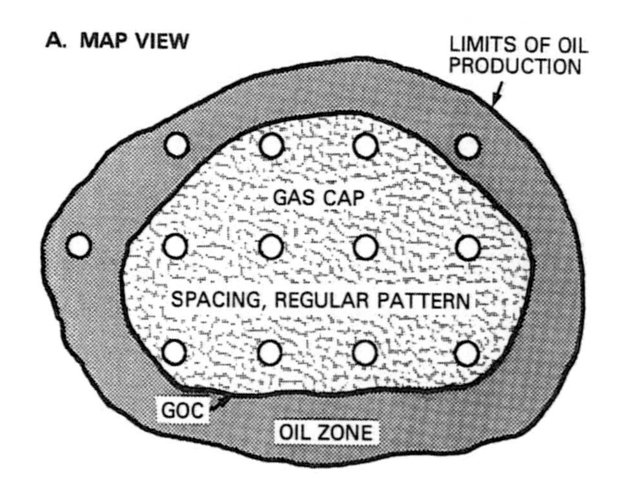

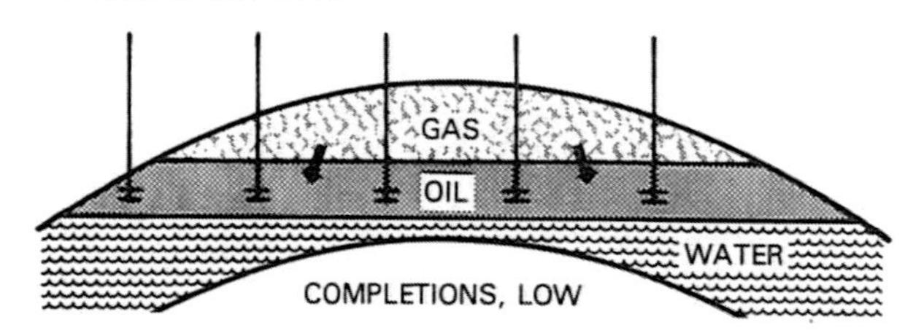

Figure 2.13. Gas-cap drive reservoir—low angle of dip. From Norman J. Clark, *Elements of Petroleum Reservoirs* (Dallas: Society of Petroleum Engineers of AIME, 1960).

The optimum spacing pattern is uniform, and wells should be completed as low in the pay section as possible. The problem in this type of reservoir is the potential for coning down the gas into the wellbore, especially at higher rates of production (fig. 2.14).

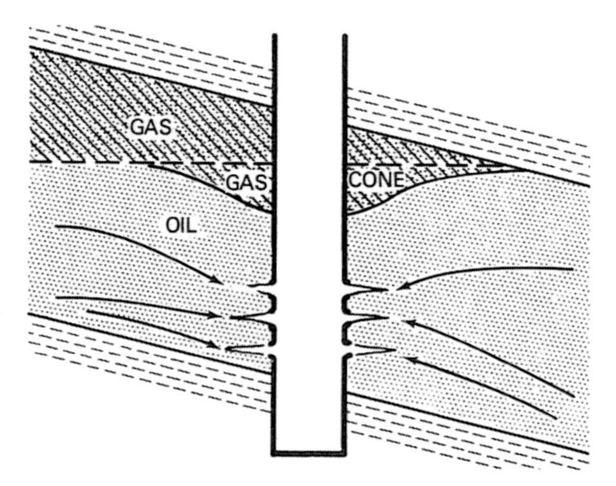

Figure 2.14. Coning (Copyright 1972 by SPE-AIME)

Coning is caused by the tendency of the gas to move vertically down through the oil column to the perforations. Usually, the goal in producing this type of reservoir is to produce only the wells making oil and shut in any high gas-oil ratio wells until the oil column is depleted. After the oil column is depleted, the gas cap is produced, or "blown down", by recompleting the wells back into the original gas-saturated interval.

Gas-cap expansion reservoir, high angle of dip. A high-relief gas-cap expansion drive reservoir presents a different type of problem. In the reservoir depicted in figure 2.15, a regular spacing pattern would result in many wells being located very close to the gas-oil contact (GOC). In this case, well locations should be made on the flank of the field where the gas column does not overlie the oil column, resulting in an irregular, oval-shaped pattern for anticline structures. Generally, wells should be completed low in the section to minimize the possibility of gas coning down from the gas cap to the perforations later in the life of the field when the gas cap has expanded downward.

Generally, wells from this type of reservoir can be produced at higher rates than those completed in a low-relief reservoir, since the possibility of coning gas down to the perforations in them is not as great. Later in the life of the field, when the gas cap expands downward, producing rates may have to be curtailed to minimize coning.

Water Drive Reservoirs

The energy for producing water drive reservoirs is derived from the expansion of the rock and water in the aquifer. Water flows into the oil column from the aquifer as the oil reservoir is depleted. Since water encroachment is part of the drive mechanism, the location and completion intervals in producing wells must take this fact into account. The angle of dip of the reservoir is an important factor affecting well-completion decisions.

Water drive reservoir, low angle of dip. A low-relief water-drive reservoir (fig. 2.16) exhibits what is called a *bottom-water drive.* The

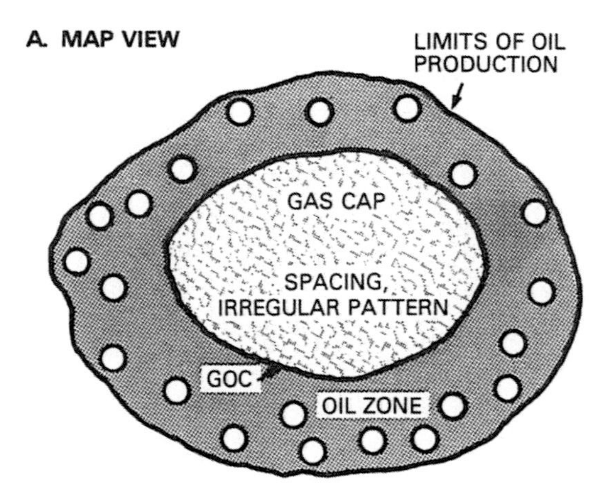

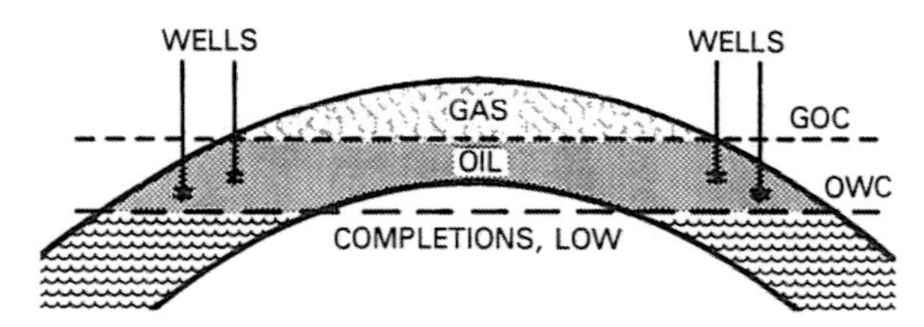

Figure 2.15. Gas-cap drive reservoir—high angle of dip. From Norman J. Clark, *Elements of Petroleum Reservoirs* (Dallas: Society of Petroleum Engineers of AIME, 1960).

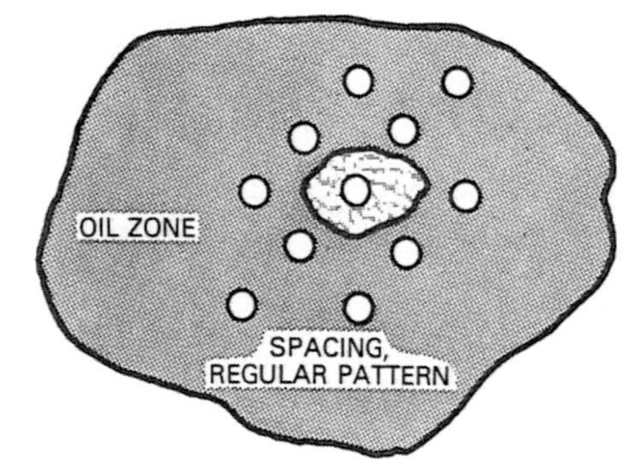

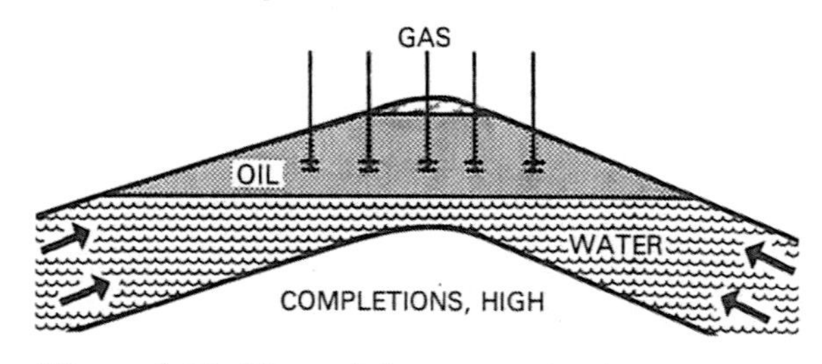

Figure 2.16. Water drive reservoir—low angle of dip, thick sand. From Norman J. Clark, *Elements of Petroleum Reservoirs* (Dallas: Society of Petroleum Engineers of AIME, 1960).

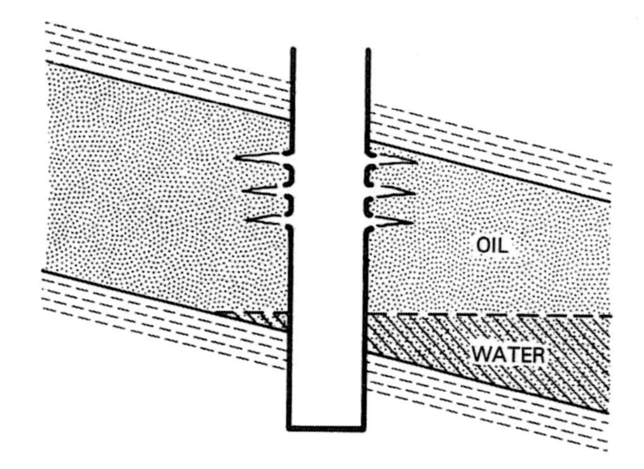

Figure 2.17. Perforated well, water drive reservoir (Copyright 1972 by SPE-AIME)

water moves up from the bottom of the reservoir as oil is produced. This situation is the reverse of that for the low-relief gas-cap reservoirs. As a result, wells are completed in a uniform pattern, but high in the section to delay the production of water (fig. 2.17). Generally, wells in low-relief water drive fields produce large volumes of water as the field approaches depletion. It is necessary to consider this fact on initial completion and to install casing of sufficient size to permit the installation of artificial lift equipment at a later date. A classic bottom-water drive oilfield is the East Texas Field.

Water drive reservoir, high angle of dip. The movement of water in a water drive reservoir with a high angle of dip (fig. 2.18) is called an *edgewater drive.* The water moves up and into the oil sand, displacing the oil in a pistonlike

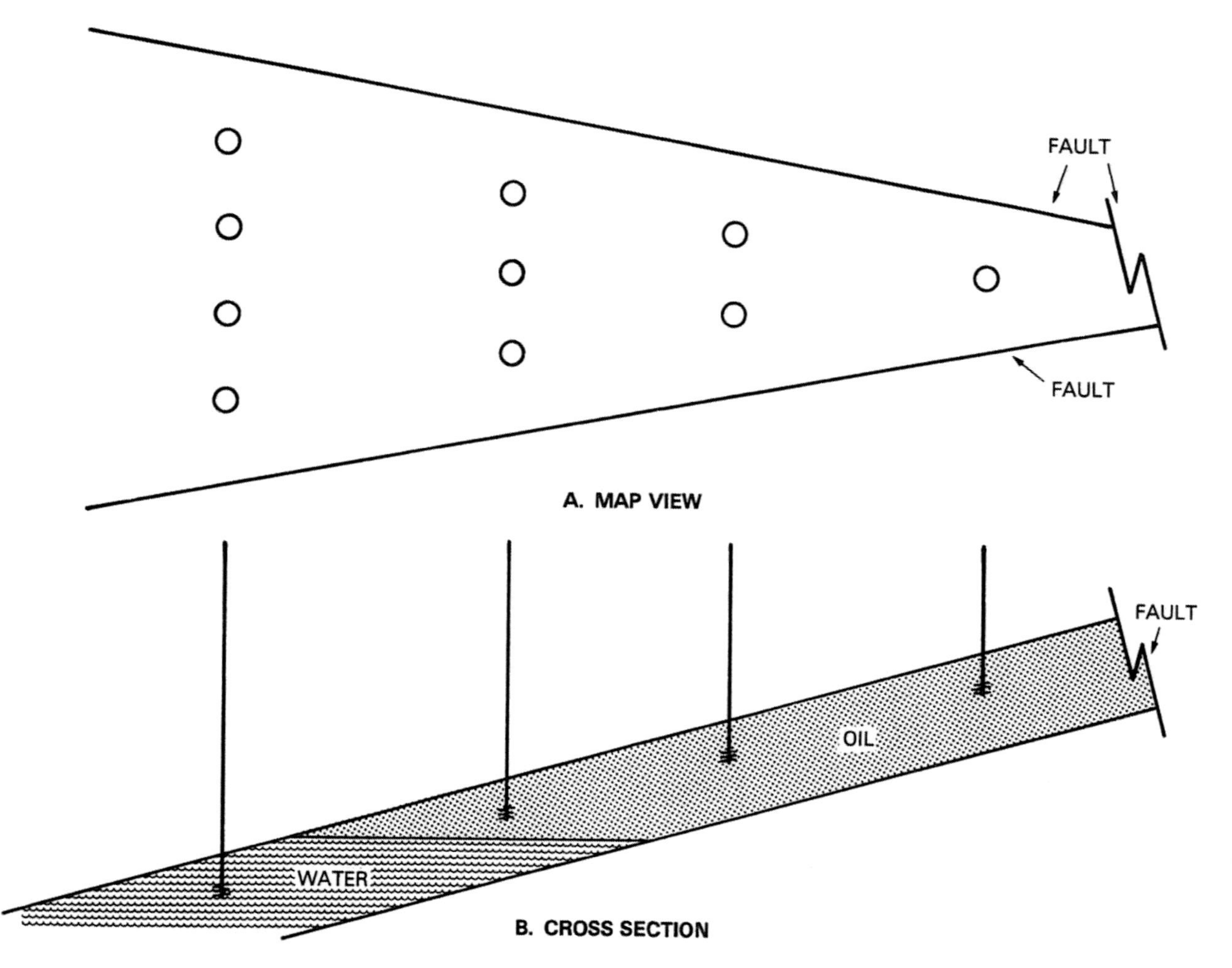

Figure 2.18. Water drive reservoir—high angle of dip

manner. It is not completely efficient, so it is sometimes referred to as a "leaky piston." Theoretically, a field of this type could be depleted by one well drilled at the top of the structure. In actual practice, more wells are needed to define the reservoir. In addition, governmental regulations or lease ownership requirements often require uniform spacing of the wells throughout the oil column. Generally, wells in this type of reservoir will flow almost to depletion and produce water for a relatively short period of time.

Wells should be completed as far updip as possible. Since the oil is being displaced in a pistonlike manner, the entire pay interval is normally perforated; there is usually no advantage to perforating only the upper part of the pay section. Unless lease ownership or governmental regulations require it, wells lower on the structure can be shut in when they start making water, since the oil will be recovered by updip wells.

Combination Drive Mechanisms, Hawkins Field Example

Quite often, reservoirs produce with a combination of the three drive mechanisms described. To illustrate this point, consider an actual field example – the Hawkins Field in East Texas (fig. 2.19). The field is an anticlinal structure caused by the intrusion of a deep-seated salt dome.

After development started in the field, some startling data were collected. The pressure in the West Reservoir was found to be 1,990 psi at 4,075 feet subsea depth. This was the pressure normally expected at this depth. Surprisingly, the pressure in the East Reservoir was measured and found to be only 1,710 psi at the same depth. This figure was very puzzling at first; the two pressures should have been comparable, since they were at the same depth in the same sand.

The Hawkins Field produces from the Woodbine Sand, a very permeable and homogeneous sandstone. It is a blanket-type sand deposit covering a large area about 80 miles by 140 miles in the East Texas Basin. The Woodbine is, of course, filled with salt water, except in the area where oil reservoirs are located.

The Hawkins Field was discovered in 1941, approximately ten years after the discovery of a number of smaller and two major Woodbine fields in the same basin, the Van and East Texas fields. By the time the Hawkins Field was discovered, large amounts of oil had been withdrawn from the Van and East Texas fields, as well as other smaller fields. Reservoir engineers finally concluded that lower pressure in the East Reservoir was due to the fact that the pressure of the Woodbine aquifer had been reduced from 1,990 psi to 1,710 psi at the Hawkins Field location as a result of production from other fields sharing this common aquifer. An asphalt seal was found to exist at the oil-water contact in the West Reservoir. This

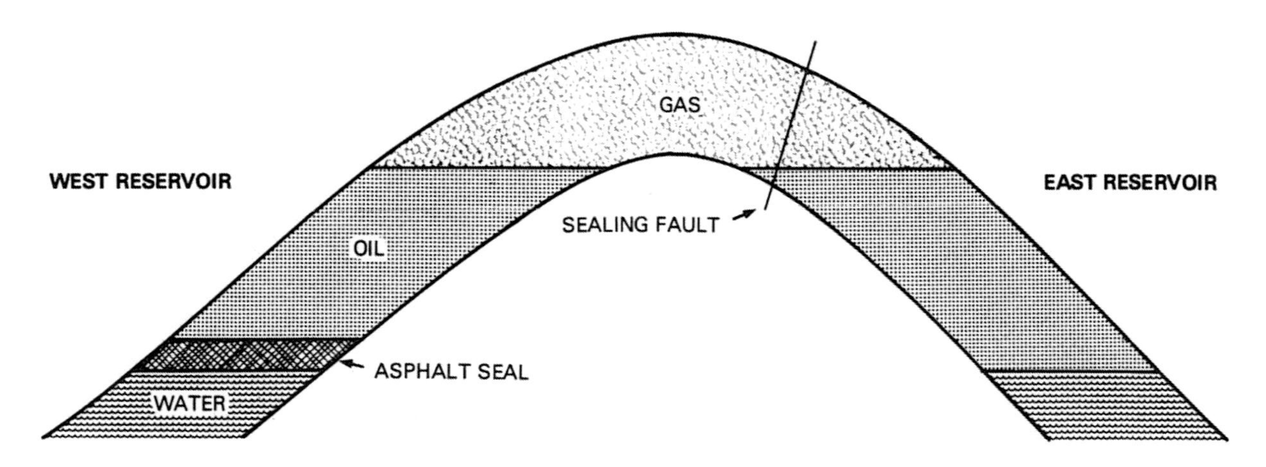

Figure 2.19. Schematic cross section of Hawkins Field in East Texas

asphalt sealed off the West Reservoir and maintained virgin pressure. The East Reservoir, which was in communication with the main Woodbine aquifer, experienced a loss of pressure along with the aquifer.

The Hawkins Field was produced for over thirty years with the West Reservoir under a gas-cap drive and the East Reservoir under a water drive. Wells were recompleted high in the section in the East Reservoir because of a bottom-water drive, and wells were recompleted low in the section in the West Reservoir because of a gas-cap drive.

The pressure in the West Reservoir decreased with continued production, increasing the differential pressure across the asphalt seal. Sometime after the field had been producing about thirty years, water production was noticed in the lowest wells on the structure. Apparently, the pressure was reduced sufficiently so that water started breaking through the asphalt seal into the West Reservoir. Once the seal had been broken, water continued to move into the reservoir, and the drive mechanism started to change from a gas-cap expansion drive to a water drive. Oil was pushed back up into the gas cap, causing a decrease in expected ultimate recovery. In order to alleviate this problem, a project was started to inject gas into the gas cap and increase the gas-cap pressure to a point where it would overcome the water influx and restore the gas-cap drive. The injection was done because ultimate recovery is expected to be higher under a gas-cap drive.

Several observations can be made from the foregoing example. First, it illustrates the point that water drives are caused by the expansion of water from closed aquifers. The expansion that forces water into oil reservoirs voided by production must necessarily result in a reduced pressure in the water in the aquifer. This situation is evidenced by the drop of approximately 280 psi at Hawkins in the Woodbine aquifer.

Secondly, the field example illustrates that the type of drive a field experiences is not fixed and can be altered as production progresses. Water will flow into an oil reservoir at a rate dependent upon the permeability of the rock and the pressure differential. It is possible to produce a potential water drive reservoir at a high enough rate to outrun the water influx so that it will pressure-deplete. This depletion is normally not desirable for an oil reservoir, but it is sometimes very desirable for gas reservoirs.

Multiple Reservoirs

Quite often a well will penetrate two or more reservoirs, creating an opportunity to make a multiple completion so that two or more reservoirs can be depleted simultaneously. The type of reservoir drives and the expected performance will determine whether a multiple completion is feasible. Artificial lift is the Achilles' heel of multiple completions, since it restricts the size of tubular goods that can be run.

No hard or fast rules can be given, but it is possible to make some general observations regarding the types of reservoirs that are suitable for multiple completions. Normally, gas reservoirs are good candidates because they can be expected to flow to depletion. Edgewater drive (high angle of dip) oil reservoirs can also be considered, because often only a few high-structure wells will require artificial lift to achieve maximum recovery. Bottom-water drive (low angle of dip) reservoirs should normally be avoided, since many wells will produce large volumes of water and require artificial lift to deplete. Dissolved-gas drive oil reservoirs of all types require artificial lift to deplete and often have secondary recovery programs such as waterflood. These two factors limit the use of multiple completions involving dissolved-gas drive oil reservoirs.

III Well Completions

Completions are basically of two types–open-hole and perforated casing. Perforated liners and tubingless completions are only variations of the perforated casing completion.

Open-Hole Completions

The first completions made in the United States were open-hole, since technology was not available to make perforated casing completions. The first commercial well in the United States was drilled in 1859; the first casing cementing job was performed in 1903; and the first perforating job was performed in 1932.

The open-hole completion continued to be used after the development of the perforated casing completion for wells that needed to be stimulated. The first limestone well was acidized in 1932, but it was not until 1950 that hydraulic fracturing became generally available for sandstone stimulation. Early open-hole completions were shot with high explosives to increase producing rates. Since shooting could be performed only in open holes, the use of open-hole completions was required.

Today, most wells are not completed open-hole, although there are still applications for this method and it should be considered. An open-hole completion is made by setting the casing at the top of the pay zone (fig. 3.1). The open-hole completion has several advantages, but the following two are the principal ones that make it still worthy of consideration:

1. Special drilling techniques and fluids can be readily applied to minimize formation damage.
2. When a gravel pack is needed, maximum productivity can be obtained from a gravel pack in an open hole, since a large-diameter screen can be run.

An open-hole completion has many disadvantages also. One of the most important is that the casing is "set in the dark" before the pay section is drilled. If the pay section happens to be poorly developed or not present, a string of casing can be wasted. Another important disadvantage is the lack of selective control over production from the open-hole section. If there are two porous intervals in the pay zone, they are commingled and cannot be produced selectively. Also, shale sections can slough off and fill up the open hole, reducing the production rate. And an obvious place where open-hole completions are not applicable is multiple reservoirs.

A variation of the open-hole completion is one with the gravel-packed screen and liner (fig. 3.2). This is the type of open-hole completion made in unconsolidated sands. It has the advantage of permitting the use of the largest size of screen and liner in a gravel-pack installation. With this type of arrangement, it is also possible to underream the formation before installing the gravel-packed screen and liner.

If the open-hole completion is made in consolidated formations, a liner may not be necessary. In some cases, a slotted liner is run to protect the open-hole section from problems with sloughing shale sections.

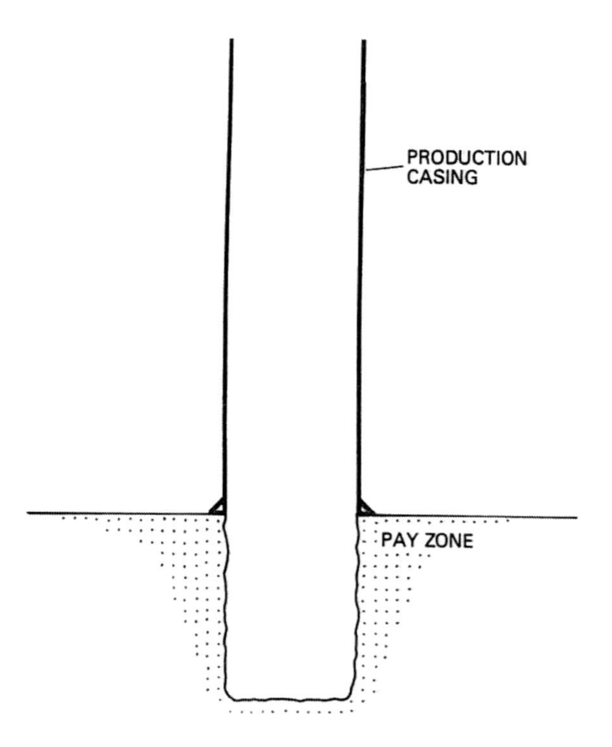

Figure 3.1. Open-hole completion (Copyright 1972 by SPE-AIME)

Perforated Casing Completions

Perforated casing completions became possible after casing cementing techniques were perfected and a method of perforating the pipe was developed. The ability to perforate and stimulate individual zones selectively makes the cased-hole completion the dominant one today. The development of logging techniques that permit an operator to evaluate potential producing intervals before setting casing gave impetus to the use of cased-hole completions (fig. 3.3). Although the entire pay interval is shown to be perforated, different intervals can be selectively perforated if desired.

In many instances unconsolidated sands can be controlled by perforated casing completions without the use of gravel packing. The Frio Sand in certain parts of the Gulf Coast area of Texas is an example of this application. The use of perforated casing completions permits the selective depletion of multipay reservoirs.

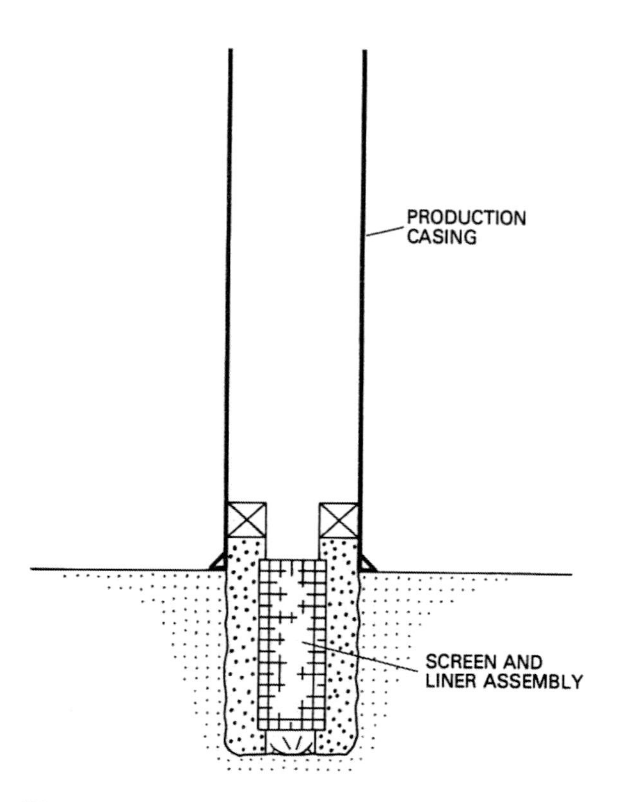

Figure 3.2. Screen and liner completion (Copyright 1972 by SPE-AIME)

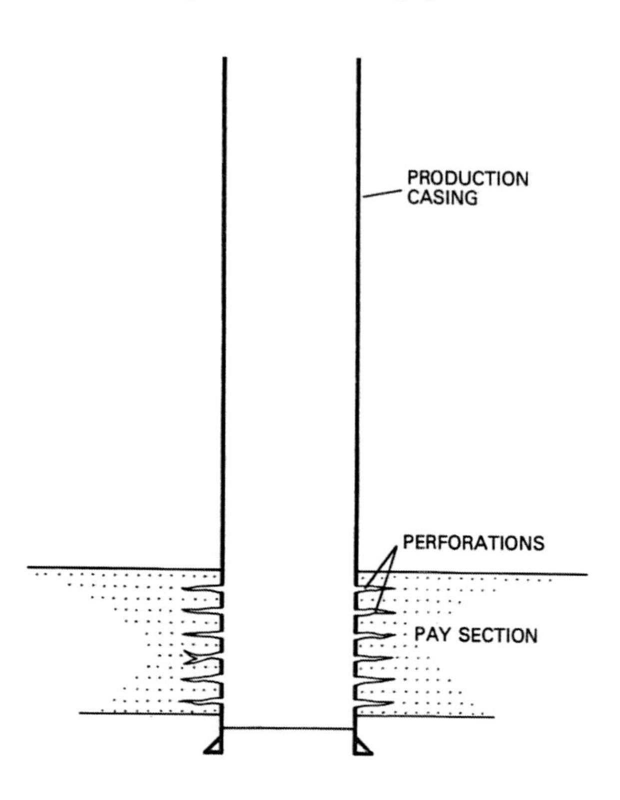

Figure 3.3. Perforated casing completion (Copyright 1972 by SPE-AIME)

Gravel-packed screen and liners can be run inside perforated casing if necessary. But since a small screen and liner is used, the productivity is less than when an open-hole screen and liner is run.

Perforated Liner Completions

The first use of perforated liners was in wells that were originally completed open-hole. The perforated liner (fig. 3.4) permits selective control of production from different intervals in the open-hole section. The success of perforated casing completions prompted operators to go back and make perforated liner completions in many open-hole completions. Initially, perforated liners were used primarily in workovers to repair existing open-hole completions.

Today, liners are widely used in the initial completion of deep wells (fig. 3.5). As well depths have increased, liners are being run to case off sections of open hole, thus eliminating the need for a full string of casing. Intermediate liners are set to case off open-hole sections that cause problems in drilling due to sloughing, lost circulation, or abnormal pressure. They are used to extend the intermediate casing string. The production liner serves as a completion casing string.

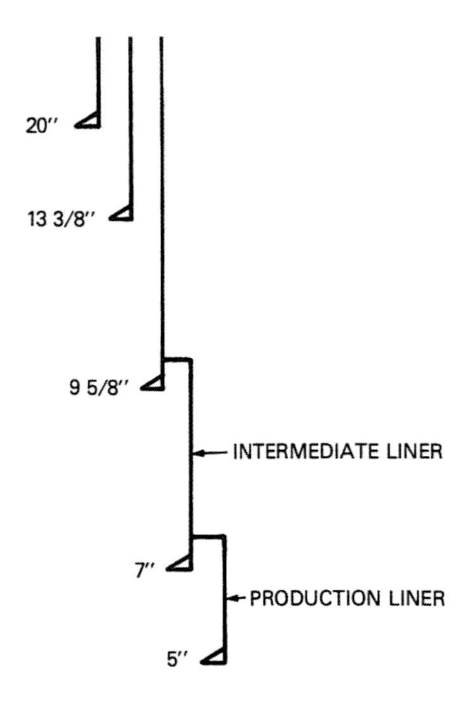

Figure 3.5. Typical liner scheme for a deep well

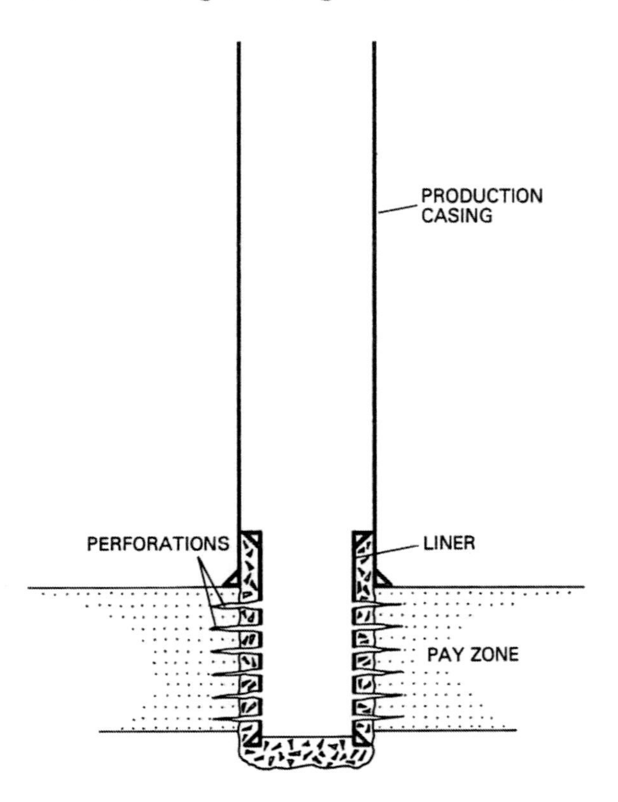

Figure 3.4. Perforated liner completion (Copyright 1972 by SPE-AIME)

The use of liners reduces the cost of completing a deep well. In addition to eliminating the need for a full string of casing, the liner can be run to depth much faster than casing, since it is run on drill pipe. Time is also saved by eliminating the need to lay down the drill pipe to pick up the casing string.

Tubingless Completions

Tubingless completions were developed in the United States in the 1950s when economic conditions were not good in the oil industry and concerted efforts were being made to reduce drilling and completion costs. In tubingless completions a smaller hole can be drilled, and the production casing string is eliminated. Tubing is run and cemented in place of casing, and the well is produced through the cemented tubing.

Completion Interval Selection

How reservoir drive mechanisms affect the well location and the completion interval selected has been shown. Selection of the correct interval will result in a more trouble-free completion and can often drastically affect the well's recovery.

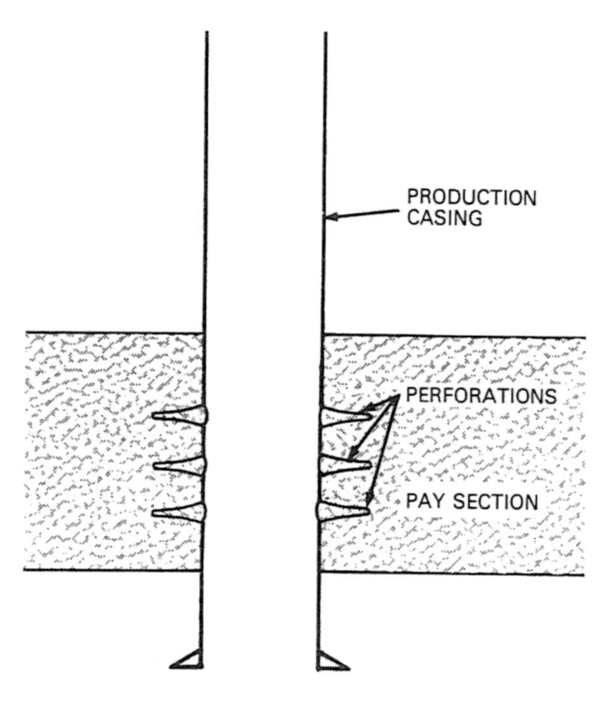

Figure 3.6. Perforating plan for a well in a competitive dissolved-gas drive field

Dissolved-Gas Drive Reservoirs

Recovery from a well in a dissolved-gas drive reservoir is dependent on its productivity as compared to total field production. In other words, each well's share of the field's reserves is based on its ability to produce. In order to ensure maximum recovery from a well in a competitive dissolved-gas drive field, it is usually desirable to perforate the entire pay section (fig. 3.6).

Some dissolved-gas drive wells may be underlain by water. Although the water will not flow into the oil reservoir at a sufficient rate to affect oil recovery from the reservoir appreciably, it can cause water production problems in an individual well. For this reason, if bottom water is present, it is generally desirable to perforate only the upper part of the pay zone.

If the well is completed in a noncompetitive field, maximum productivity may not be required. In this situation, it is desirable to perforate only as much pay as is required to achieve the desired producing rate.

If the well is completed in a very thick reservoir with high permeability, gravity segregation may occur so that a secondary gas cap will form at the top of the pay section. Wells completed at the top of the section will produce with a high gas-oil ratio. Some Middle East fields are 800 to 1,000 feet thick. For this type of reservoir, it may be desirable to perforate only the bottom part of the pay so that completion will not be made in the area where a secondary gas cap might form.

Water Drive Reservoirs

The two basic types of water drive reservoirs are bottom-water and edgewater drives. In bottom-water drive, the water-oil contact moves up from the bottom at a rate determined by upstructure field withdrawals. In this type of situation, it is desirable to perforate only the upper part of the pay zone to delay the production of water by the well (fig. 3.7).

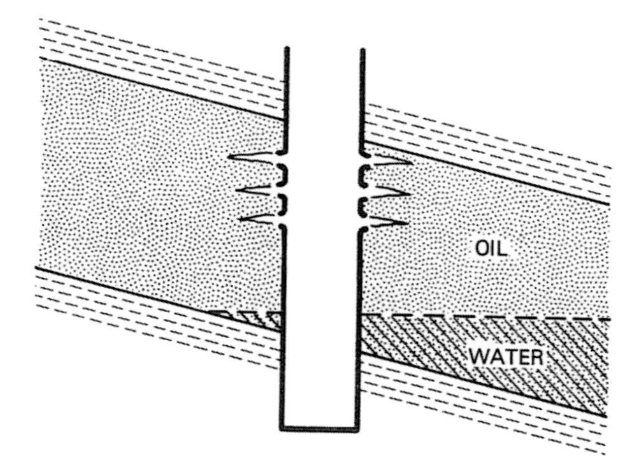

Figure 3.7. Perforated well—water drive reservoir (Copyright 1972 by SPE-AIME)

A classic example of a bottom-water drive reservoir is the north end of the East Texas Oil Field. The field produces from the Woodbine Sand, which is a very homogeneous sandstone with an average permeability of 2,500 to 3,500 millidarcys. In the 1950s and 1960s, the water-oil contact moved up about 1 to 1½ feet per year. By perforating only the top foot of pay, water production could be delayed until the oil column at the wellbore had been displaced. Because of the high permeability, the perforation of 1 foot of the pay section resulted in sufficient productivity. For example, if a well had 25 feet of pay section above the water-oil contact, water production could be delayed about fifteen to twenty years by perforating only the top 1 foot.

The edgewater drive mechanism is often piston-like, so the oil column at the wellbore is displaced by water in a very short time after the water reaches the wellbore. This situation is typical of many high-relief, Frio Sand reservoirs on the Gulf Coast. In this case, perforating a limited interval at the top of the pay does not materially affect well recovery. Since the Frio Sand is unconsolidated, it is usually desirable to perforate a large section of the pay zone to limit the pressure drawdown through each perforation, thus minimizing sand production problems.

Gas-Cap Expansion Drive Reservoirs

The gas-cap expansion field is the reverse of the bottom-water drive field. The gas-oil contact moves downward with production, and the oil column at the well is eventually displaced by gas. To prevent premature gas production and to prolong the oil-producing life of the well, it is usually desirable to perforate the bottom part of the pay section (fig. 3.8).

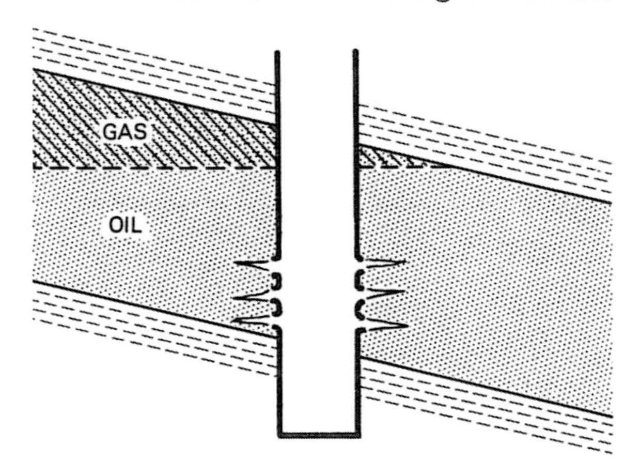

Figure 3.8. Perforated well—gas-cap expansion drive reservoir (Copyright 1972 by SPE-AIME)

Unknown Drive Mechanism Reservoirs

The reservoir drive mechanism cannot always be determined immediately, and the drive mechanism may be unknown when some of the early wells in a field are drilled. A well may have a potential for either a gas-cap expansion drive or a bottom-water drive (fig. 3.9).

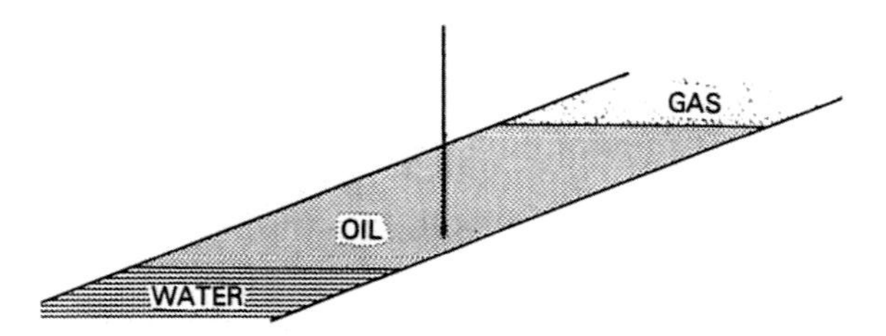

Figure 3.9. Well with unknown drive mechanism

It is usually desirable in the situation shown to start out by perforating the bottom part of the pay section first. If a gas-cap expansion drive develops, the completion interval is correct. If a water drive develops, the well will start making water when the water-oil contact reaches the bottom of the hole. If it does, the lower perforations can be squeezed off or isolated by a bridge plug and the well reperforated in the top of the pay section. This procedure is a cheap workover with a high rate of success. If the upper part of the pay section were perforated first and a gas-cap drive developed, premature gas breakthrough would occur. In order to repair the breakthrough, it would be necessary to squeeze off the perforations and drill out. This type of workover is more expensive, with a high failure rate. Also, it is very undesirable to have squeezed-off gas perforations above the oil-producing interval, since the perforations may break down and threaten a blowout on future workover operations.

Types of Completion Arrangements

Originally, all well completions were of the open-hole type, usually utilizing one tubing string for production. Technology rapidly developed to meet the new conditions of higher pressures, multiple reservoirs, and other problems as wells became successively deeper.

Single Completions

Single completions can be divided into two classifications – those with and those without tubing strings.

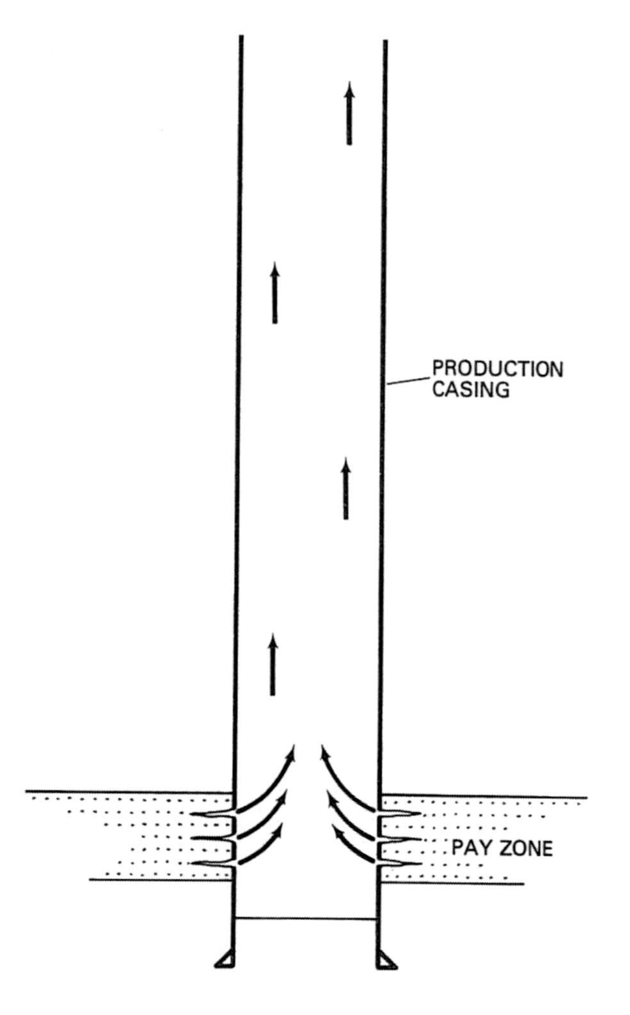

Figure 3.10. Single completion without tubing (Copyright 1972 by SPE-AIME)

Single completion without tubing. A single completion without tubing is the simplest type, since production flows up the production casing (fig. 3.10). Although a perforated casing completion is shown, a perforated liner or an open-hole completion would also be applicable.

This completion method has the advantage of simplicity, since there is no complicated downhole equipment to fail. If large-diameter casing is run, it is possible to produce the well at high rates. For example, rates of 30,000 to 40,000 barrels per day could be produced by this arrangement.

The single completion without tubing has some serious disadvantages, however, and as a result it is seldom used. The disadvantages are –

1. There are no means for artificial lift if the well stops flowing.
2. If the well fluids are corrosive, casing corrosion will occur, and injecting corrosion inhibitors is very difficult.
3. The only way the well can be killed is by *bullheading*. In bullheading, the well is overpowered by pumping a kill fluid down the casing. Some wells can be damaged by this practice.
4. Casing is subjected to well pressure, and a casing leak may result in loss of fluid to the formation.

Single completion with tubing. The disadvantages to the single completion without tubing can be overcome by running a tubing string (fig. 3.11). The addition of a tubing string allows the well to be produced by artificial lift when it stops flowing. In addition, flow through the smaller-diameter tubing will prolong the flowing life of the well. The well can be allowed to flow up the annulus and the tubing string simultaneously if a high rate of flow is desired. Periodically, a corrosion inhibitor can be pumped down the tubing string to protect the casing against corrosion.

The tubing can be used as a kill string if it becomes necessary to kill the well in order to perform a workover. The kill fluid can be pumped down the annulus and circulated back out the tubing string, thus eliminating the necessity to bullhead fluids into the formation.

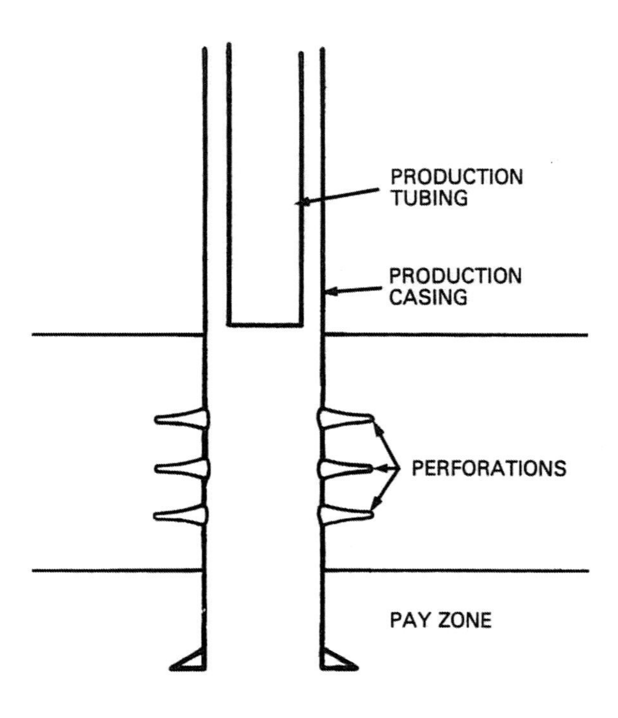

Figure 3.11. Single completion with tubing string

Dual Completions

Further development in well completion technology brought about dual completion techniques. Sometimes it is desirable to produce two zones simultaneously, but sound reservoir management requires that the two zones be produced separately. The single-packer dual completion was the first arrangement designed to solve this problem.

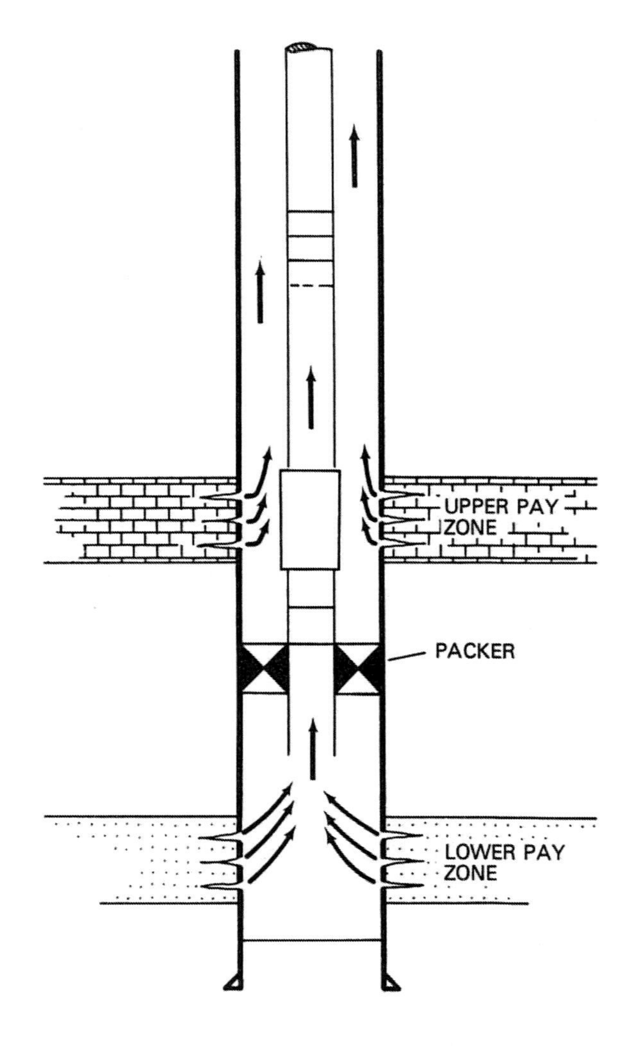

Figure 3.12. Dual completion with single packer (Copyright 1972 by SPE-AIME)

Dual completion with single packer. The single completion with tubing shown in figure 3.11 was converted to a dual completion simply by adding a packer (fig. 3.12). This dual completion had the advantage of simplicity and low cost, and many installations were made. One basic problem, however, was annular flow. The upper zone flowed through the tubing-casing annulus, presenting no problem if the upper zone was gas, but presenting many problems if the upper zone produced oil.

Oil flow through an annular area can be a problem, especially if the well tends to load up and die. The annular area cannot be swabbed to restore production. Also, if the upper zone produces corrosive fluids, it is difficult to inject corrosion inhibitors. The arrangement is satisfactory only if the upper zone is noncorrosive dry gas.

Artificial lift becomes a problem whenever either zone quits flowing. The lower zone can be lifted through the tubing string, but with difficulty. The upper zone cannot in any practical way be artificially lifted through the annulus.

Alleviation of one of the problems with annular flow was accomplished with the crossover packer.

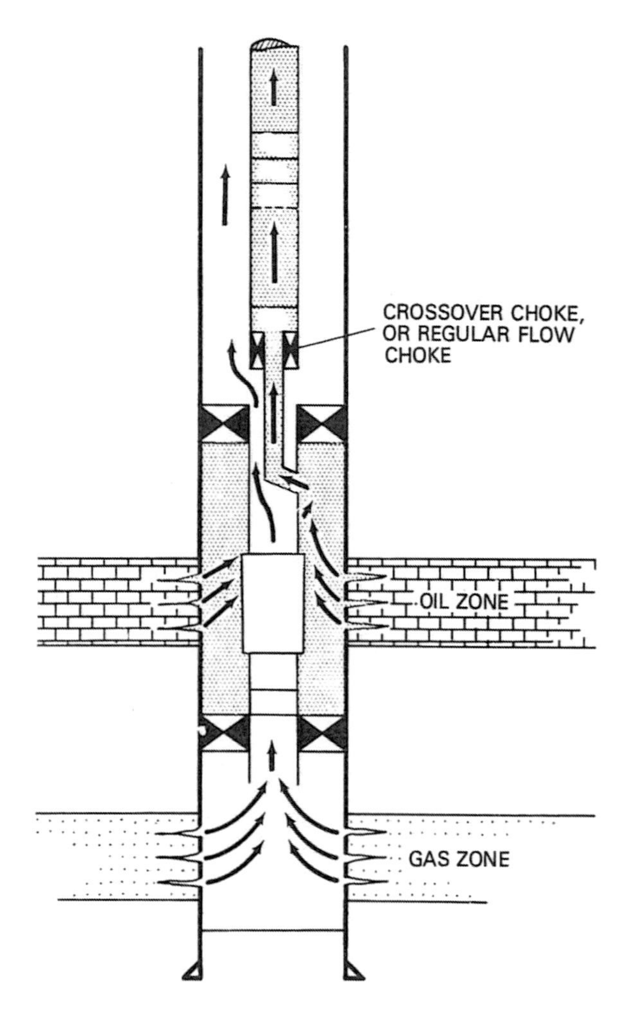

Figure 3.13. Dual completion with crossover packer (Copyright 1972 by SPE-AIME)

Dual completion with crossover packer. The dual-completion single-packer arrangement is satisfactory for one gas zone and one oil zone, provided the gas zone is on top. For those cases in which the gas zone is on the bottom, the crossover packer was developed (fig. 3.13).

Although the crossover packer solved the problem for dual completions in which the gas zone is on the bottom, it did not help with the artificial lift or annular flow problem when the upper zone was oil. The next step was the development of the dual completion with two parallel tubing strings.

Dual completion with two parallel tubing strings. The previously considered dual-completion arrangements have the problem of annular flow and present difficult and sometimes impossible artificial lift problems. In the early 1950s, a dual completion utilizing two parallel strings of tubing was developed (fig. 3.14). Since both zones are produced up tubing strings, the annular flow problem is eliminated. Also, the artificial lift problem is made less severe, although there are still problems in lifting dual wells with parallel strings of tubing. Venting gas from the lower zone can present a problem in wells pumped with sucker rods.

Triple Completions

As wells became deeper and more expensive to drill, technology continued to improve, and

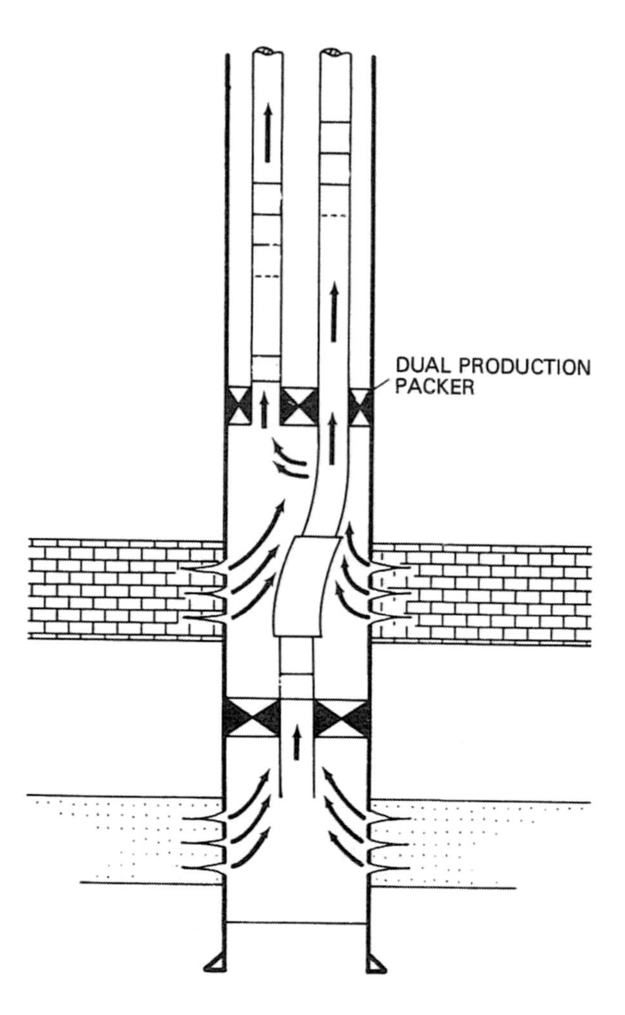

Figure 3.14. Dual completion with two parallel tubing strings (Copyright 1972 by SPE-AIME)

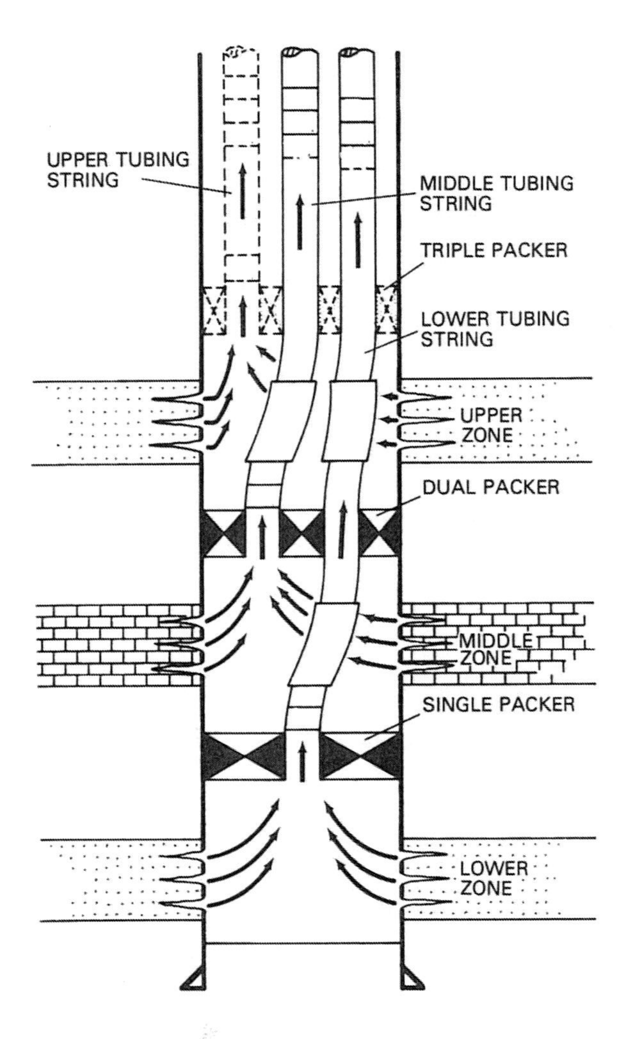

Figure 3.15. Triple completion (Copyright 1972 by SPE-AIME)

triple completions (fig. 3.15) became possible for multipay fields. The triple completion permits the simultaneous depletion of three producing intervals and for this reason has a great economic advantage. With three strings of tubing, the upper tubing string can be eliminated if the upper zone is suitable for annular flow.

Although triple completions permit three zones to be produced simultaneously, they are very complex and susceptible to communication problems. The economics of each installation has to be considered; there are instances in which the risk of equipment failure can be accepted due to the more rapid payout of the well by the increased producing rate.

Alternate Completions

Improving technology made possible the design of alternate completions. Sometimes two or more productive intervals are encountered in a well, but it is desirable to produce them one at a time. For example, there may be two gas zones capable of producing 5 MMcf/d each, and pipeline capacity is only 4 MMcf/d. Also, proper reservoir management requires that the zones be produced separately. The problem can be solved with an alternate completion.

Single completion with alternate. If only two zones are present, the single completion with alternate (fig. 3.16) can be made. The lower

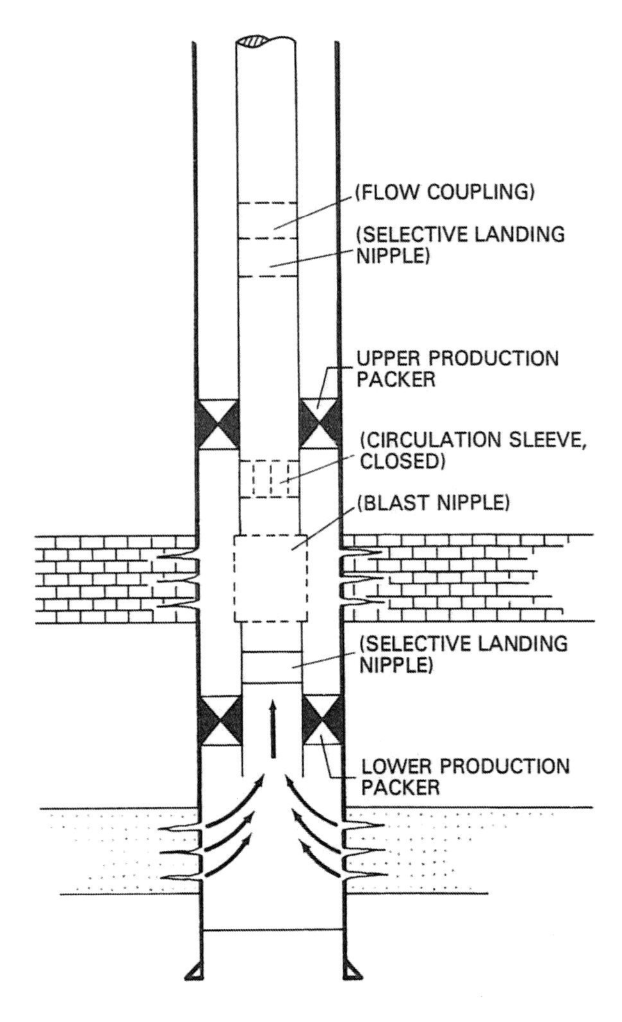

Figure 3.16. Single well with alternate completion (Copyright 1972 by SPE-AIME)

zone in the figure is being produced up the tubing, and the circulation sleeve is closed, shutting in the upper zone. As soon as the lower zone is depleted, it can be blanked off by running a plug on wireline and setting it in the selective landing nipple located immediately above the lower packer. The upper zone is then placed on production by using a wireline tool to open the circulating sleeve immediately below the upper packer. Production can be shifted from one zone to another by the use of a wireline, thus eliminating a costly workover, especially important in offshore or marshy areas where the costs of conventional workovers are very high.

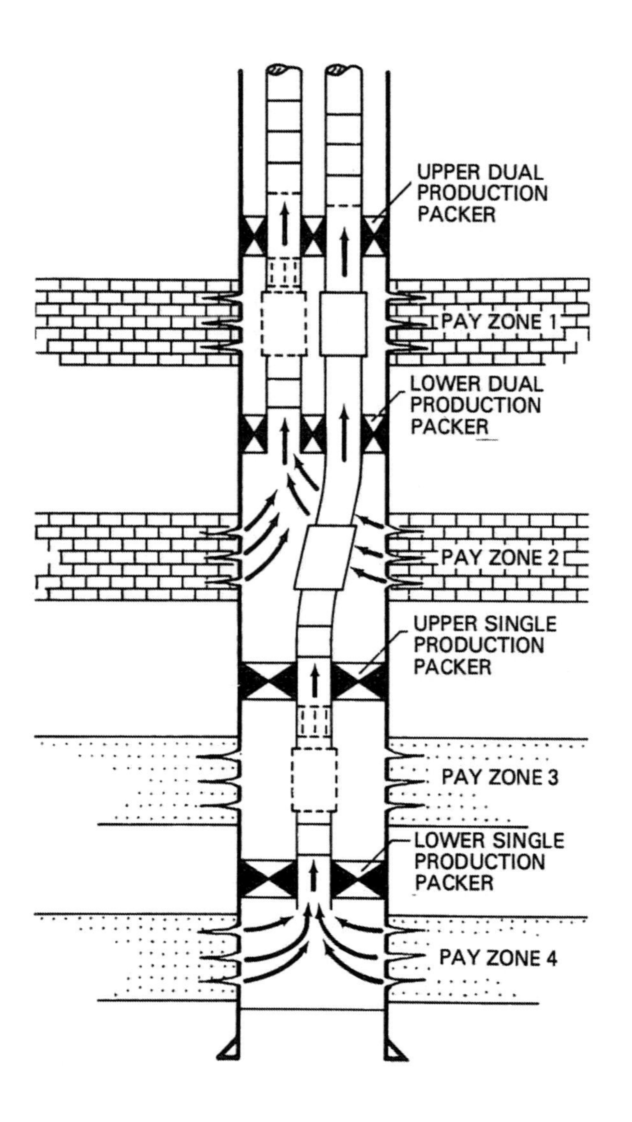

Figure 3.17. Dual well with two alternate completions (Copyright 1972 by SPE-AIME)

Dual well with two alternate completions. Alternate completions can also be applied to dual completions (fig. 3.17). The arrangement is similar to the single well with an alternate installation (fig. 3.16). If the producing intervals are numbered 1 through 4, starting at the top, zones 2 and 4 are being produced first. Production can be shifted to zones 1 and 3 by running blanking plugs and opening circulating sleeves with a wireline unit. The switch to zones 1 and 3 can be made individually.

The main disadvantage to this arrangement is the complexity of the downhole equipment. This installation has four packers and two sliding sleeves, increasing the chance for a costly equipment failure. The risk may be justified in marshy or offshore areas where workover costs are prohibitively high.

Tubingless Completions

Tubingless completions present difficult artificial lift and workover problems due to the smaller size of the production tube. There are instances in which they are applicable, and they should still be considered, especially in gas reservoirs. Recently, some triple tubingless completions utilizing 3½-inch OD tubing were made on gas wells in the Gulf of Mexico.

Single tubingless completion. The tubingless completion for a flowing well shown in figure 3.18 is identical to the conventional casing single completion depicted in figure 3.10, except that tubing is used for casing. In the United States, casing normally refers to pipe of 4½" OD and larger. A tubingless completion is one in which 2⅜" OD, 2⅞" OD, or 3½" OD tubing is run for casing. Tubingless completions can be artificially lifted, although the volumes that can be lifted are limited. Either gas lift or beam pumping can be used.

Multiple tubingless completions. Multiple completions in which two to four strings of tubing are run have been made. Cementing tech-

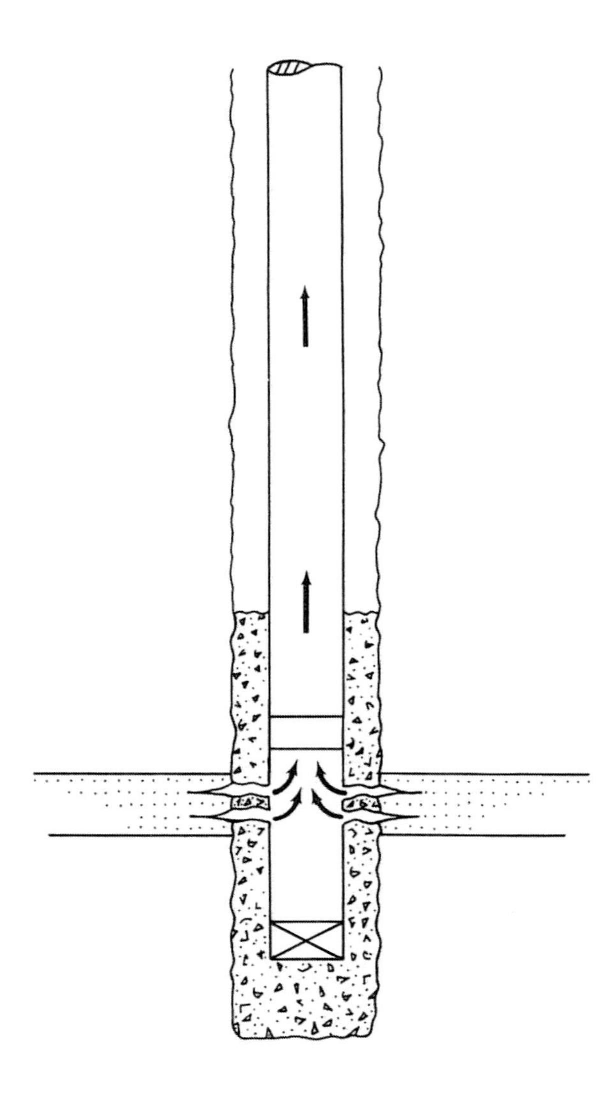

Figure 3.18. Single tubingless completion (Copyright 1972 by SPE-AIME)

niques have now been improved so that the possibility of achieving a good cement job on multiple completions has been greatly improved. A triple tubingless completion is shown in figure 3.19. Multiple tubingless completions offer a possible solution to the problem of rapidly depleting multipay fields where the size of the reserves gives rise to marginal economics. They are especially attractive for gas fields where artificial lift is not a problem.

The triple tubingless completion is the one that offers a unique advantage when high flow rates are desired in a multipay field. Three strings of 3½" OD tubing can be run, thus permitting high flow rates. A conventional triple completion with 3½" OD tubing would require 10¾" OD production casing. A big savings in casing costs can be realized. In addition, conventional triple completions require costly, complex downhole equipment (fig. 3.15). This equipment is also eliminated by using a tubingless completion.

In water drive gas fields, a tubingless completion using 3½" OD tubing is applicable. It has been noted that recoveries from water drive gas fields can be improved from about 60% of the

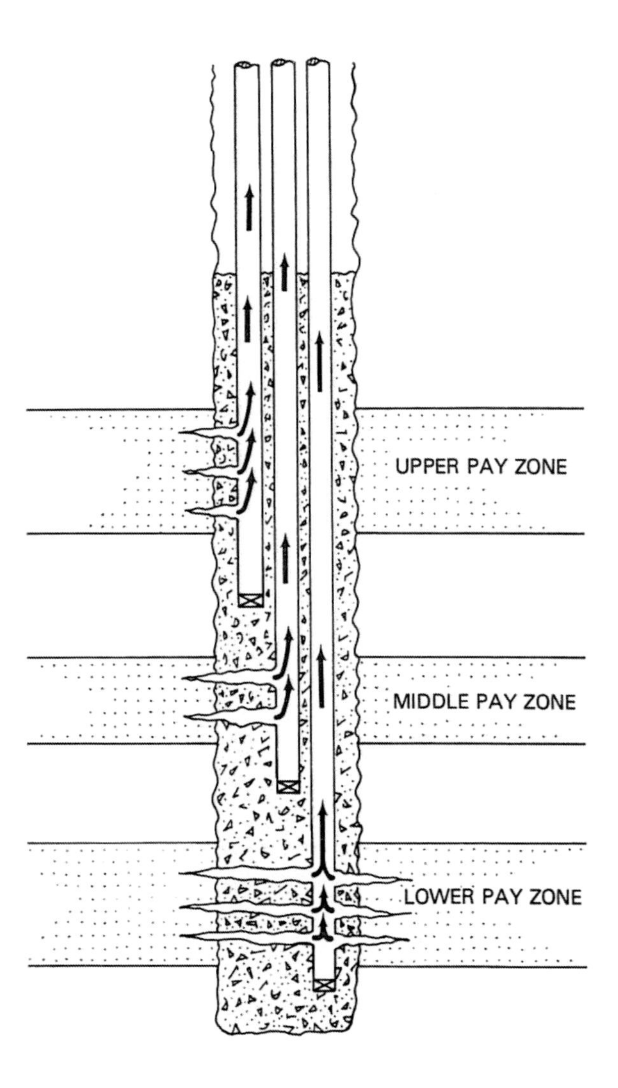

Figure 3.19. Triple tubingless completion (Copyright 1972 by SPE-AIME)

gas in place to 70% or 80% by producing the field at a rate high enough to outrun water encroachment. This rate requires large-sized tubing. For example, dry gas will flow at a rate of approximately 20 MMcf/d through 2⅞" OD tubing from a 6,000-foot well with a flowing surface pressure of 500 psi and a flowing bottomhole pressure of 1,600 psi. With 3½" OD tubing, it will flow at about 35 MMcf/d under the same conditions.

Triple 3½" OD tubingless completions have been run from depths of 4,000 feet to 13,000 feet, with the major application being in the 4,000- to 6,000-foot range.

IV Evaluating Well Performance

One of the primary functions of the production engineer is to evaluate and predict well performance. After a well is completed, periodic well tests show how much oil, gas, and water the well makes. In order to deplete the well intelligently, it is necessary to predict whether it will require artificial lift and at what date. The production engineer makes these predictions primarily on the basis of periodic well tests and flowing and shut-in bottomhole pressure data. Basically two questions require answers: (1) What is the well producing now? and (2) What will it produce if producing conditions are changed? To arrive at an answer to the first question, periodic well tests are run.

Periodic Well Tests

As soon as a well is completed and placed on production, most operators start periodic well tests. Although some wells are produced into separate production facilities with their production determined daily, most wells produce into common production facilities along with other wells. For wells that produce into common facilities, it is necessary to run periodic well tests to determine how much each well is contributing to the total production.

Periodic tests were originally performed manually, and many wells are still tested in this manner. Since automated production facilities have become more common, automatic well testing is becoming the dominant method because it usually results in more frequent testing. Regardless of the method, the same data are usually collected. A typical oilwell test illustrates the type of data collected.

Date: 7/1/83

Oil production = 750 BOPD
Water production = 150 BWPD
Gas rate = 450 Mcf/d
Gas-oil ratio = 600 cu ft/bbl
Flowing tubing pressure = 900 psig

A typical gas well test might contain the following data:

Date: 7/1/83

Gas rate = 16.55 MMcf/d
Condensate production = 695 BCPD
Water production = 5 BWPD
Flowing tubing pressure = 2,500 psig
Bbl condensate/MMcf gas = 42

Although the data collected varies, the above tests show what is typically recorded on periodic well tests. A review of periodic well tests can be used to monitor well and reservoir performance.

The first test after completion verifies the type of fluids the well will produce. If it is a discovery well, the type of production to expect, based on log interpretation, may not be certain, since some wells are difficult to evaluate by logs. The first test will confirm the type of fluid production and the rate at which the well produces.

Periodic tests are usually taken on a routine basis, and the results are plotted in graphic form to permit rapid evaluation. Most companies store well test data in computers and have periodic printouts for operating personnel to use.

Periodic well tests also indicate the type of reservoir drive mechanism that is operating. The flow rate and gas-oil ratio should remain fairly constant for a water drive well. Although available subsurface data may indicate in advance the type of drive mechanism to expect, nothing is sure until production actually begins. If there are a rather constant oil rate, flowing bottomhole pressure, and gas-oil ratio, they will tend to confirm the existence of a water drive mechanism. On the other hand, if there are a decreasing oil rate and constantly increasing GORs, a dissolved-gas drive is indicated.

Periodic well test data can also be used in problem well analysis to determine whether a well has production problems that require remedial action.

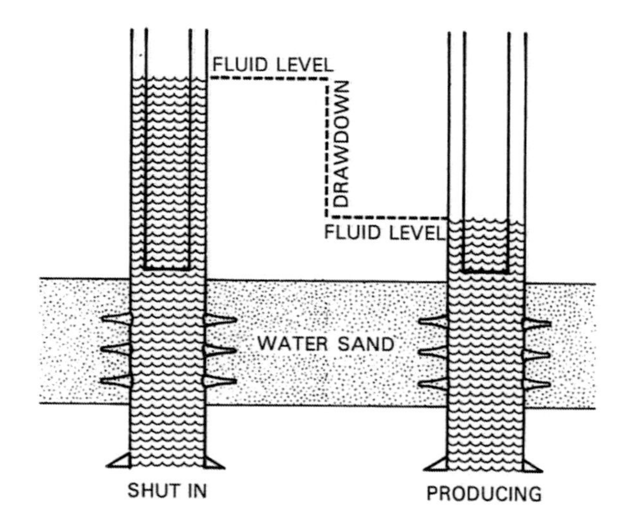

Figure 4.1. Water-well productivity evaluation

Productivity Index Method

The productivity index (PI) method was developed to answer the question, What will the well produce if conditions are changed? Periodic well tests show what a well will produce under current conditions. Quite often it is necessary to predict what the well will produce if producing conditions are changed. For example, it may be important to predict the production increase if artificial lift is installed. The first method used by production engineers to predict well productivity was the productivity index method. The PI method was originally used to evaluate water wells, and the method was adopted by production engineers to evaluate oilwells.

Water wells are evaluated in the following manner. First, the well is shut in, and the distance to the static water level is measured (fig. 4.1, *left*). The well is then produced, and the distance to the producing water level is measured (fig. 4.1, *right*). The difference between the two water levels is called the *drawdown*. In the case of a water well, the drawdown is measured in feet. If a well produced 5,000 gallons of water per day with a drawdown of 100 feet, it would be expected to produce 10,000 gallons per day if the drawdown were increased to 200 feet. A straight-line relationship is assumed, another way of saying that the producing rate is proportional to the drawdown.

The productivity index (PI) for the example water well can be expressed as follows:

$$PI = \frac{5{,}000 \text{ gal/day}}{100 \text{ feet of drawdown}}$$

$$PI = 50 \text{ gal/day/foot of drawdown.}$$

In other words, for every foot the fluid level is decreased, the flow rate will increase 50 gallons per day. If the distance from the static fluid level to the producing sand is 300 feet, then the maximum drawdown that can be achieved is 300 feet. Since the well will produce 50 gallons per day for every foot the fluid level is lowered, the maximum rate at which the well will produce is–

$$\text{Maximum rate} = 50 \text{ gal/day/foot of drawdown} \times 300 \text{ feet}$$

$$\text{Maximum rate} = 15{,}000 \text{ gal/day.}$$

Although water well drawdown is normally expressed in feet, it actually expresses the pressure at the producing sand. The pressure exerted by a 1-foot column of water is 0.433 psi/ft. The static bottomhole pressure in the above example is 129.9 psi (300 ft × 0.433 psi/ft = 129.9 psi.)

The flow calculation made above can be expressed as an equation, as follows:

p_e = static reservoir pressure, psi

p_w = producing bottomhole pressure, psi

q = flow rate, gal/day.

$$PI = \frac{q}{p_e - p_w}$$

$$PI = \frac{5{,}000}{129.9 - 86.6} = 115.5 \text{ gal/day/psi drawdown.}$$

The producing rate at a drawdown of 200 feet is calculated as follows:

Producing fluid level = 300 − 200 = 100 ft above perforations

Producing wellbore pressure = 100 × 0.433 = 43.3 psi

$$PI = \frac{q}{p_e - p_w}$$

$$115.5 = \frac{q}{129.9 - 43.3}$$

$$q = 115.5\,(129.9 - 43.3)$$

$$q = 10{,}000 \text{ gal/day with 200 feet of drawdown.}$$

It is necessary to express drawdown in terms of pressure when evaluating oilwells, since the weight of different oils varies. In addition, the amount of gas in solution will affect the weight of oil. Since oil production is usually expressed in barrels/day, the oilfield units for *PI* and *q* are as follows:

q = flow rate, barrels/day

PI = barrels/day/psi drawdown.

The following example will illustrate the use of the productivity index in evaluating an oilwell. Assume that a well has been tested with the following results:

q = 300 BOPD

p_w = 1,500 psi

p_e = 1,800 psi

It is known from the above test that the well produces 300 BOPD with a producing wellbore pressure of 1,500 psi and a static reservoir pressure of 1,800 psi. If the producing rate at a p_w of 1,000 psi is desired, it can be determined by using the PI method.

$$PI = \frac{300 \text{ BOPD}}{1{,}800 - 1{,}500} = 1.0 \text{ BOPD/psi drawdown}$$

The flow rate at p_w = 1,000 psi can be found as follows:

$$q = PI\,(p_e - p_w) = 1.0\,(1{,}800 - 1{,}000) = 800 \text{ BOPD.}$$

Another question that might be asked is: What is the maximum producing rate for the well? The maximum producing rate q_m will occur at a p_w = 0 psi.

$$q_m = 1.0(1{,}800 - 0) = 1{,}800 \text{ BOPD.}$$

The maximum rate possible from this well is 1,800 BOPD based on the PI method. A maximum rate is a theoretical number because p_w can't be lowered to zero, but p_w can approach zero with artificial lift, so it serves as a good tool in comparing well productivities.

It is obvious from the above example that if a well is tested at one rate, and the static reservoir pressure (p_e) and the producing wellbore pressure (p_w) are obtained, it is possible to determine the rate at which the well will produce at any other value of p_w. This calculation is a very powerful tool in evaluating well performance.

The PI method is very useful because it permits the evaluation of a well's producing capability without the necessity for actually testing it at a higher rate. In order to test at a higher rate, it may be necessary to install larger tubing or equip the well for artificial lift. The PI method permits an estimate of the higher rate without the expense of changing equipment. The producing rate estimated by the PI method can often be used to justify the equipment changes.

After discussing the mechanics of using the PI method, it is appropriate to determine when it is technically sound to use the PI method in well performance evaluation. The factors affecting fluid flow in the reservoir should first be reviewed. Darcy's law describes radial flow for an oilwell as follows:

$$q = \frac{7.08\, k_o h\, (p_e - p_w)}{B\mu\, ln\, r_e/r_w} \quad \text{(Darcy's law)}$$

In oilfield units–

q = flow rate, barrels/day

k_o = permeability to oil, darcys

h = pay thickness, ft

B = formation volume factor, vol/vol

μ = viscosity, cp

r_e = drainage radius, ft

r_w = wellbore radius, ft

p_e = reservoir pressure at drainage boundary, psi

p_w = wellbore flowing pressure, psi.

The equation demonstrates that the flow rate is directly proportional to $p_e - p_w$, which represents the pressure drawdown. If the other factors remain constant, then the producing rate will increase as the drawdown is increased. Since the reservoir pressure can't be changed, the drawdown is increased by lowering the wellbore pressure.

If all of the data necessary to solve Darcy's equation, such as k, h, B, μ, r_e, and r_w are available, it is possible to calculate the flow rate at different wellbore pressures (p_w). Unfortunately, all of these data are not normally available, and the productivity index is relied on to determine what a well will produce at different conditions.

The PI equation and Darcy's law can be combined, as follows:

$$PI = \frac{q}{p_e - p_w}$$

$$q = \frac{7.08\, k_o h\, (p_e - p_w)}{B\mu\, ln\, r_e/r_w}$$

$$PI = \frac{\dfrac{7.08\, k_o h\, (p_e - p_w)}{B\mu\, ln\, r_e/r_w}}{(p_e - p_w)}$$

Simplifying,

$$PI = \frac{7.08\, k_o h}{B\mu\, ln\, r_e/r_w}$$

The above expression can be simplified further for estimation purposes. For common values of r_e and r_w, the expression $ln\, r_e/r_w$ varies between a value of 7 and 8. It is not too sensitive to changes in r_e or r_w, since the natural logarithm of the ratio is used in the equation. In view of this,

$$PI \cong \frac{k_o h}{B\mu}$$

By making the above approximation, the expression for *PI* is simplified to only four terms. The approximation is useful in several ways. It illustrates how changes in k, h, B, and μ will affect the PI. The approximation can be used to estimate a PI for a well before it is tested.

The following example illustrates how to use the approximate PI to evaluate a new well. Assume that a new well has just been drilled in a Texas Gulf Coast Frio Sand. From offset wells, it is known that the following data are typical:

k_o = 450 md = 0.45 darcys

μ = 0.63 cp

B = 1.24 vol/vol

From the electric log of the new well, the pay thickness is estimated to be 22 feet.

The PI of the new well can be estimated as follows:

$$PI \cong \frac{k_o h}{B\mu} = \frac{0.45 \times 22}{1.24 \times 0.63} = 13 \text{ BOPD/psi.}$$

The approximate-PI method gives a reasonable estimate of the PI before a well is drilled. The PI of a new well can also be estimated by using the PIs of offset wells adjusted for differences in net pay. For example, if an offset well has a PI of 14.4 BOPD/psi and a net pay of 25 feet, then the specific PI = 14.4/25 = 0.58 BOPD/psi/ft.

The PI for the new well is estimated as follows:

$$PI \text{ (new well)} = 22 \text{ ft} \times 0.58 \text{ BOPD/psi/ft} = 13 \text{ BOPD/psi.}$$

The PI method was the only one available to evaluate well performance until 1968. It has many limitations and is often used where it is not applicable.

Difficulties with PI Method

It has been pointed out that the following approximate relationship can be developed to express PI:

$$PI \cong \frac{k_o h}{B\mu}$$

The PI relationship assumes a straight-line, or linear, relationship for the productivity of a well at higher rates. In other words, if the pressure drawdown is doubled, the producing rate will be doubled (fig. 4.2). Unfortunately, a straight-line relationship is not observed on many wells. After the PI method was used to evaluate oil-wells, engineers began to notice that in dissolved-gas reservoirs, PIs declined when the wells were produced at higher rates and also as the reservoir was pressure-depleted. They found that instead of a straight-line relationship, wells producing from dissolved-gas reservoirs exhibited a curved relationship (fig. 4.3).

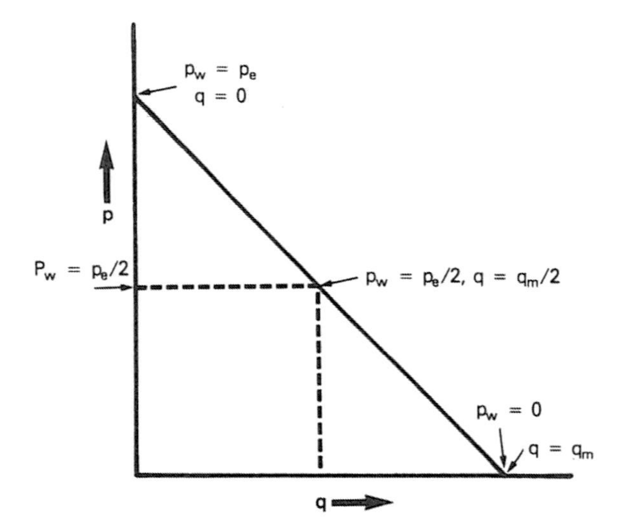

Figure 4.2. Linear PI

The figure shows that the actual producing rate at higher drawdowns is less than would be predicted by the productivity index. The maximum drawdown occurs when $p_w = 0$. Notice the difference between the actual and the predicted maximum producing rates. A review of the approximate-PI equation will show why this occurs.

$$PI \cong \frac{k_o h}{B\mu}$$

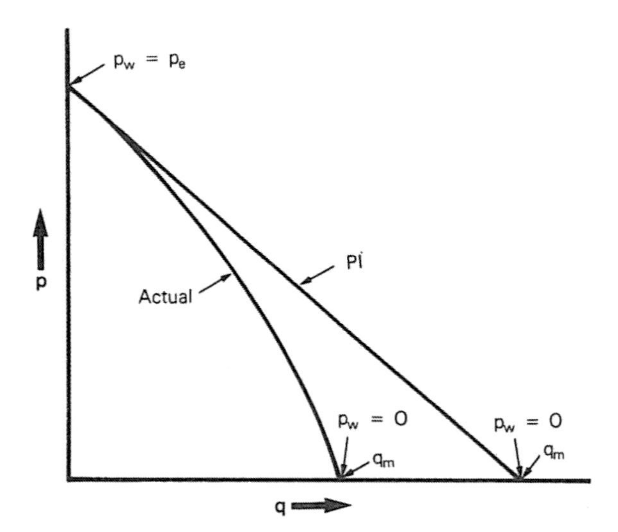

Figure 4.3. Nonlinear PI

It was pointed out in chapter 1 that k, B, and μ are all pressure-dependent properties of the crude oil. As the pressure in an oil reservoir is decreased, changes take place in all of the above variables that affect the PI of the well. What happens to k, B, and μ as the pressure decreases below the bubble point should be briefly reviewed.

k_o: The relative permeability to oil decreases as the pressure drops, and gas comes out of solution and builds up a gas saturation.

B: The FVF decreases as gas comes out of solution with decreased pressure.

μ: The reservoir oil viscosity increases as the gas comes out of solution with decreased pressure.

A decrease in the FVF will tend to increase the flow rate, while a decrease in relative permeability and increase in oil viscosity will decrease the flow rate. The relative permeability change has a much greater effect on flow rates as the pressure drops and gas breaks out of solution. The net effect of the changes in the three variables is a drop in the PI of the well as the wellbore producing pressure decreases.

Dissolved-gas reservoirs, discussed in chapter 2, pressure-deplete with production. As a result, the PIs for a dissolved-gas reservoir start to decline as soon as the reservoir pressure drops below the bubble point, and they continue to decline until the reservoir is pressure-depleted.

The normal reservoir pressure decline in a dissolved-gas reservoir changing k_o, B, and μ accounts for the decline of PIs with time. It is still necessary to consider the decline of PI that occurs at a point in time at higher rates of production (fig. 4.3).

The decline in PI at higher rates of production is illustrated in figure 4.4. The pressure distribution for a well in radial flow is shown. Because of the radial flow effect, the pressure drops sharply as the wellbore is approached. The pressure in the area near the wellbore is much lower than it is at the drainage boundary. Curve q_1 represents the pressure distribution in the reservoir at the first rate of production, with a wellbore pressure p_{w1}.

If the well is produced at a higher rate, q_2, it is necessary to decrease the pressure at the wellbore to p_{w2}. Decreasing the pressure at the wellbore will cause the pressure distribution curve to drop downward.

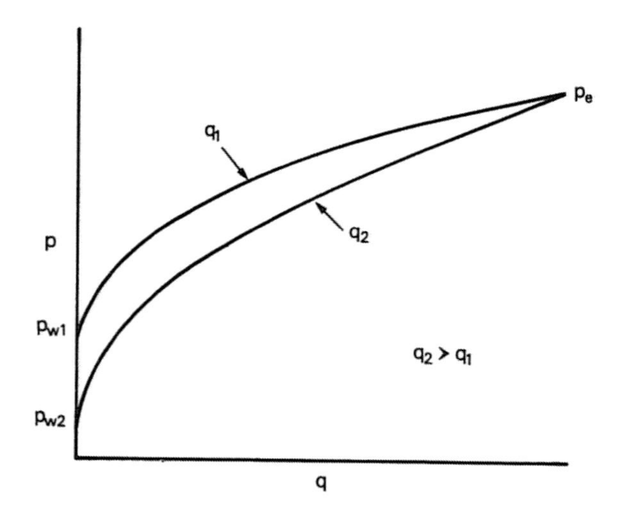

Figure 4.4. Pressure distribution at higher flow rates

The gas saturation near the wellbore will increase because of the lower pressure at the higher rate, q_2. The oil viscosity will also increase near the wellbore because more gas comes out of solution at the lower pressure. The extra free gas competes with the oil for flow space, so the relative permeability to oil will be reduced. The net effect will be a decrease in the producing rate, or PI, over what would be expected by a straight-line PI relationship.

The PI method is therefore not applicable to dissolved-gas reservoirs below the bubble point. In order to evaluate inflow performance of dissolved-gas drive wells below the bubble point, the Vogel equation must be used.

Vogel's Equation

The PI method was used to evaluate dissolved-gas drive wells for a long period of time even though many people recognized that the relationship was not always valid. Muskat pointed out in 1941 that the PI method is not valid when both oil and gas flow in the reservoir. The PI method was used for lack of a better method until 1968, when J. V. Vogel proposed a new approach.

Vogel made a computer study in 1968 of the performance of dissolved-gas drive reservoirs below the bubble point. As a result of his investigation, he found that the inflow performance of an oilwell producing from a dissolved-gas drive reservoir below the bubble point can be expressed by the following equation:

$$\frac{q_o}{(q_o)_{\max}} = 1 - 0.2\left(\frac{p_{wf}}{p_r}\right) - 0.8\left(\frac{p_{wf}}{p_r}\right)^2$$

where

q_o = oil flow rate, BOPD at p_{wf}
$(q_o)_{max}$ = maximum flow rate, BOPD @ $p_{wf} = 0$
p_{wf} = wellbore flowing pressure, psi
p_r = static reservoir pressure, psi.

It can be seen from figure 4.5 that Vogel has plotted p_{wf}/p_r vs. $q_o/(q_o)_{max}$. If a well test complete with flowing bottomhole pressure and

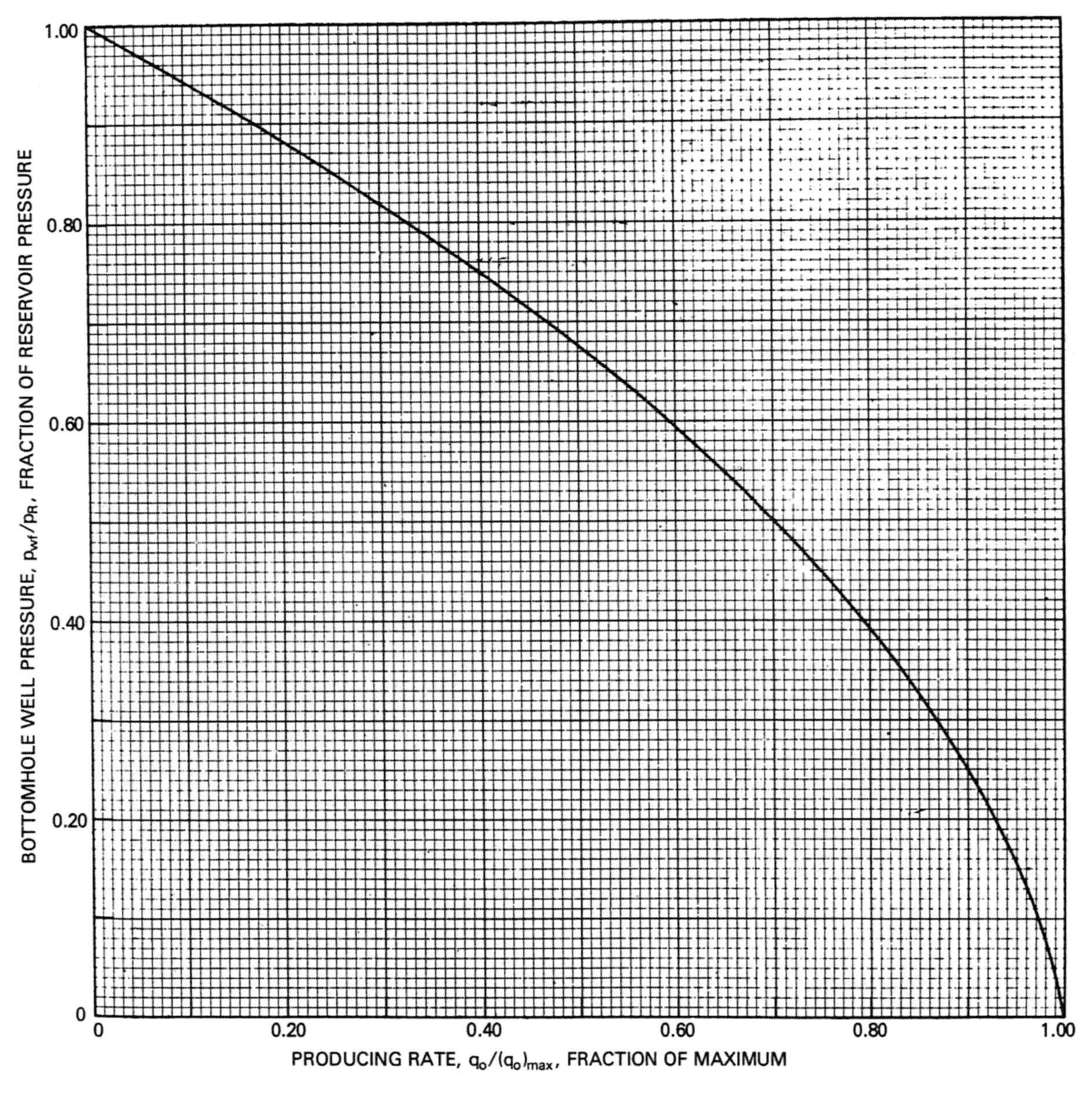

Figure 4.5. Inflow performance relationship for solution-gas drive reservoirs. From J. W. Vogel, "Inflow Performance for Solution-Gas Drive Reservoirs," ***Journal of Petroleum Technology,*** **January 1968, 83.**

static reservoir pressure are available, the performance of a dissolved-gas drive well can be evaluated below the bubble point pressure by the Vogel method.

The following example will illustrate the use of the Vogel method in evaluating a dissolved-gas drive oilwell. The same data previously used in the example problem for the productivity index calculation will be used to illustrate the importance of using the correct method. The following well data apply:

$$q_o = 300 \text{ BOPD}$$
$$p_{wf} = 1{,}500 \text{ psi}$$
$$p_r = 1{,}800 \text{ psi.}$$

What will the well produce if p_{wf} is lowered to 1,000 psi?

As has been noted, Vogel has plotted p_{wf}/p_r vs. $q_o/(q_o)_{max}$ (fig. 4.5). From the well test data, p_{wf}/p_r can be calculated as follows:

$$\frac{p_{wf}}{p_r} = \frac{1{,}500 \text{ psi}}{1{,}800 \text{ psi}} = 0.83.$$

The value of 0.83 indicates that on the test the wellbore flowing pressure was decreased to a value equal to 83% of the static reservoir pressure. The value of p_{wf}/p_r indicates how much pressure drawdown is required to produce the test rate. This value of p_{wf}/p_r can be used to determine what value of $q_o/(q_o)_{max}$ goes with $p_{wf}/p_r = 0.83$. From the Vogel chart (fig. 4.6), the value of $q_o/(q_o)_{max}$ can be read.

The value of $q_o/(q_o)_{max} = 0.28$ should be read. Since the value of q_o is known, $(q_o)_{max}$ can now be calculated as follows:

$$\frac{q_o}{(q_o)_{max}} = 0.28$$

$$(q_o)_{max} = \frac{q_o}{0.28}$$

Since $q_o = 300$ BOPD,

$$(q_o)_{max} = \frac{300}{0.28} = 1{,}070 \text{ BOPD.}$$

The maximum rate, or $(q_o)_{max}$, for the example well is 1,070 BOPD. A maximum rate of 1,800 BOPD was calculated using the PI method. For a dissolved-gas drive reservoir, the estimate of maximum producing rate is 68% too high if it is calculated with the PI method.

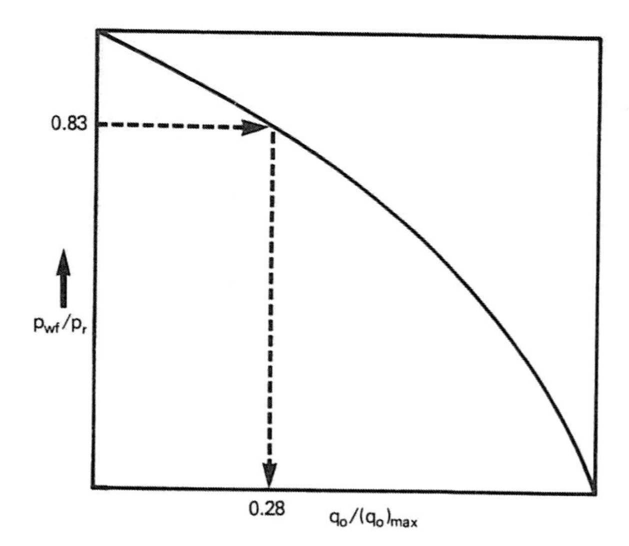

Figure 4.6. A reading from the Vogel chart

After $(q_o)_{max}$ is known, production of the well if $p_{wf} = 1{,}000$ psi can be evaluated. First a new p_w/p_r is calculated.

$$\frac{p_{wf}}{p_r} = \frac{1{,}000}{1{,}800} = 0.56$$

Entering the Vogel curve in figure 4.5 with a $p_{wf}/p_r = 0.56$, a value of 0.64 is obtained for $q_o/(q_o)_{max}$.

$$\frac{q_o}{(q_o)_{max}} = 0.64 \text{ at } \frac{p_{wf}}{p_r} \; 0.56$$

Since the value of $(q_o)_{max}$ has already been calculated to be 1,070 BOPD, the value of q_o @ $p_{wf} = 1{,}000$ psi can be calculated as follows:

$$\frac{q_o}{1{,}070} = 0.64$$

$$q_o = 0.64 \times 1{,}070 = 685 \text{ BOPD.}$$

When this well was evaluated by the PI method, a producing rate of 800 BOPD at $p_{wf} = 1{,}000$ psi was calculated.

It can be seen from the above example that the error will be greater at increased drawdowns. By the PI method, a rate 68% too high was calculated at maximum drawdown when $p_{wf} = 0$. At a $p_{wf} = 1{,}000$ psi, the producing rate of 800 BOPD determined by the PI method is 17% greater than calculated by the Vogel method.

The example problem can also be solved by using Vogel's equation:

$$\frac{q_o}{(q_o)_{max}} = 1 - 0.2 \frac{(p_{wf})}{(p_r)} - 0.8 \frac{(p_{wf})^2}{(p_r)}$$

If the Vogel equation is used to solve for the flow rate q_o at a new value of p_{wf}, the result is a straightforward calculation. If the new value of p_{wf} is solved for at an assumed value of q_o, it is necessary to solve a quadratic equation. For this reason, it is a little more convenient to use the chart.

The following example illustrates the use of the Vogel method to solve for a new value of p_{wf} at a new value of q_o. Assume that it is desired to know how much p_{wf} will have to be reduced in order to double the rate.

Well Test Data

q_o = 300 BOPD
p_{wf} = 1,500 psi
p_r = 1,800 psi
$(q_o)_{max}$ = 1,070 BOPD

Assuming that the test rate will be doubled, q_o = 600 BOPD.

$$\frac{q_o}{(q_o)_{max}} = \frac{600}{1{,}070} = 0.56.$$

Using the Vogel chart (fig. 4.5) and a $q_o/(q_o)_{max}$ value of 0.56, a value of 0.63 is read for p_w/p_r.

Since p_r = 1,800 psi,
p_{wf}/p_r = 0.63
p_{wf} = 0.63 × 1,800 = 1,130 psi at a rate of 600 BOPD.

Selecting the Correct Evaluation Method

The performance of a well producing from a strong water drive reservoir is obviously going to be different from one producing from a dissolved-gas drive reservoir. Selection of an evaluation method for well performance should take the type of reservoir into consideration.

Dissolved-Gas Drive Reservoirs

It has been noted that wells producing from dissolved-gas drive reservoirs above the bubble point have constant gas-oil ratios. The wells produce at essentially the solution gas-oil ratio for the crude. Since gas is not breaking out of solution and interfering with the flow of oil to the wellbore, the PI method is applicable.

Below the bubble point pressure, gas breaks out of solution in the reservoir and impedes the flow of oil to the wellbore. To account for the phenomenon, the Vogel method should be used.

Water Drive Reservoirs

Wells producing from water drive reservoirs have essentially constant producing gas-oil ratios, and the reservoir pressure is maintained by water influx. No problem is experienced with gas breaking out of solution, so the PI method can be used. The important thing to remember is that water influx into a reservoir is dependent upon the size of the aquifer and the permeability of the reservoir rock. Each reservoir has a maximum rate at which it can be produced without exceeding the water influx rate. If the water influx rate is exceeded, then the reservoir will perform like a dissolved-gas drive reservoir. Vogel's equation will then be applicable.

Gas-Cap Expansion Drive Reservoirs

Wells producing from the oil column below the gas-oil contact in a gas-cap expansion reservoir produce with a constant gas-oil ratio. This type of well can be evaluated by using the PI method.

After gas breaks through to a well, neither the PI nor the Vogel method is applicable. In most instances, the wells should be shut in after the gas cap breaks through in order to conserve gas-cap energy to produce down-structure wells.

General

Engineers adopted the PI method from water well testing, but dissolved gas reservoirs below the bubble point don't perform like water wells because there is gas in solution in the oil. When the pressure in the reservoir drops below the bubble point pressure, gas comes out of solution and impedes the flow of oil to the wellbore, making the PI method inappropriate. In order to overcome this problem, the Vogel method was developed. Following is a summary of the application of the PI and Vogel methods to different types of reservoir drives.

PI Method

1. Use to evaluate dissolved gas-drive wells above the bubble point.
2. Use to evaluate gas-cap expansion wells prior to gas-cap breakthrough.
3. Use to evaluate water drive wells.

Vogel Method

1. Use to evaluate dissolved-gas drive wells below the bubble point.

V Primary Cementing

The Drake well, the first commercial oilwell drilled in the United States, was completed in 1859. It was not until 1903 that the first U.S. use of a cement slurry in an oilwell was made in the Lompoc Field in California. Initially, the cement slurry was just dumped down the hole after the casing had been run. The dump method of casing cementing was replaced in 1910 by the two-plug method. In the two-plug method, slurry is pumped down the casing and into the annulus between the pipe and the hole. One plug is used to isolate the slurry from the mud being displaced inside the casing, and the second plug to isolate the slurry from the displacing fluid. The two-plug system is the method used today to cement casing strings.

It can be seen from this brief historical outline that the use of cements in oilwells is a fairly recent innovation. The oilwell cementing process has become very sophisticated, and it is now possible to cement wells satisfactorily under a wide variety of conditions.

Cement slurry properties are tailor-made by the use of additives to fit the unique conditions of each cement job. Designing a proper slurry is one of the prerequisites of a good primary cement job. Selecting cements and additives requires consideration of many factors.

Types of Cement

The first wells were cemented by using ordinary portland cements obtained from lumber yards. These cements were satisfactory when the wells were shallow and the bottomhole temperatures were low. As the well depths increased, with a resultant increase in bottomhole temperature, new cements had to be developed to meet the new conditions.

One of the first problems encountered with the use of ordinary portland cement in oilwells was the time required for it to set up. This was especially critical on shallow surface casing strings. At the low surface temperatures encountered in most surface casing settings, ordinary portland cement might take 24 to 36 hours to set up. This caused a delay in drilling operations. In order to overcome this problem, a high early-strength cement was developed by making a finer grind than was used on ordinary cements. This type of cement was designated Type C by the American Petroleum Institute (API).

As well depths increased with a corresponding increase in temperature and pressure, ordinary cements set up too fast, and there was not sufficient time to place them properly. To overcome this problem, API Classes D, E, and F cements were developed. These cements have longer pumping times at higher temperatures and pressures as a result of having a coarser grind and a change in the composition of the cement.

Another problem that was encountered was cement deterioration from contact with sulfate waters in some formations. To help alleviate this problem, sulfate-resistant cements were developed by limiting the amount of tricalcium aluminate in the cement.

The first response in supplying cements to meet the more exacting conditions as wells were drilled deeper was to develop new special-application cements. By 1968, the API listed the following six cements for oilwell use.

Class A: Ordinary portland good for wells to a depth of 6,000 feet.

Class B: Similar to Class A, except that it is available in both moderately and highly sulfate-resistant types. Good to 6,000 feet.

Class C: High early strength achieved by making a finer grind. Good to 6,000 feet. Available in moderately and highly sulfate-resistant types.

Class D: Good for wells of 6,000 to 10,000 feet. Available in moderately and highly sulfate-resistant types.

Class E: Good for wells of 10,000 to 14,000 feet. Available in moderately and highly sulfate-resistant types.

Class F: Good for wells of 10,000 to 16,000 feet and at extremely high pressures and temperatures. Available in highly sulfate-resistant types.

The six cements listed above satisfied the requirements for most wells, but they presented a stocking problem. With the advent of bulk storage, the problem of stocking as many as six different cements at a single location presented a problem. The use of several cements at offshore drilling locations presented severe logistical problems. To solve this problem, the cement industry and the API developed the basic cement concept in 1968. In the basic cement concept, only one basic cement is stocked, and its properties are altered by the addition of additives to enable it to be used at any depth. The API approved the following two basic cements:

Class G: Good as basic cement to 8,000 feet, and to deeper depths with additives. Available in moderately and highly sulfate-resistant types. Worldwide cement.

Class H: Good as basic cement to 8,000 feet. Can be run deeper with additives. Available only in moderately sulfate-resistant types.

The adoption of the basic cement concept allows all cementing requirements on a well to be met with one cement. Class G is becoming the worldwide standard.

Additives

The adoption of the basic cement concept necessitated the use of various additives to obtain the desired cement slurry properties. Present-day casing cementing applications vary from below freezing conditions in the permafrost areas of Alaska and Canada to temperatures of 450° F to 500° F in steam wells. The following sections discuss the most widely used additives and their application.

Accelerators

The time required for the cement to set up is dependent primarily on the temperature. Although the temperature increases with depth, the temperature at the bottom of most surface casing strings is usually rather low, since these casing strings are set at shallow depths. As a result, the waiting on cement (WOC) time can be very long. One way to get around this problem is to use the Class C high early-strength cement. The reduction in WOC time is usually accomplished by adding an accelerator to Class A, B, G, or H cements. The most commonly used accelerator is calcium chloride ($CaCl_2$). The effect of $CaCl_2$ on thickening times of Class A cements is given in table 5.1. Note that the thickening time for casing cementing with Class A cement at 2,000 feet can be reduced from 4 hours and 12 minutes to 1 hour and 43 minutes by the addition of 2% $CaCl_2$.

The table also lists the compressive strengths of Class A cement after curing for different lengths of time at different temperatures. For example, the Class A cement with 2% $CaCl_2$ will have a compressive strength of 1,170 psi after being cured 6 hours at 95° F and 800 psi. The same Class A without $CaCl_2$ will have a compressive strength of only 235 psi when cured for 6 hours at 95° F and 800 psi.

Sodium chloride (NaCl) in low concentrations can also be used as an accelerator. Data on its use as an accelerator are given in table 5.2. In higher concentrations, NaCl also acts as a retarder.

Retarders

The problem with surface holes is that ordinary cement does not set up fast enough

TABLE 5.1
EFFECT OF CALCIUM CHLORIDE UPON THE THICKENING TIME AND COMPRESSIVE STRENGTH OF API CLASS A CEMENT

	Thickening Time (hours: min.)							
	API Casing Cementing Tests for Simulated Well Depth (ft) of				API Squeeze Cementing Tests for Simulated Well Depth (ft) of			
Calcium Chloride (percent)	1,000	2,000	4,000	6,000	1,000	2,000	4,000	6,000
0.0	4:40	4:12	2:30	2:25	3:30	3:29	1:52	0:58
2.0	1:55	1:43	1:26	1:10	1:30	1:20	0:54	0:30
4.0	0:50	0:52	0:50	0:58	0:48	0:53	0:37	0:23

		Compressive Strength (psi)				
		At Atmospheric Pressure, and Temperature of			At API Curing Pressure and Temperature of	
Curing Time (hours)	Calcium Chloride (percent)	40°F	60°F	80°F	800 psi 95°F	1,600 psi 110°F
6	0	N.S.	20	75	235	860
12	0	N.S.	70	405	1,065	1,525
24	0	30	940	1,930	2,710	3,680
48	0	505	2,110	3,920	4,820	5,280
6	2	N.S.	460	850	1,170	1,700
12	2	65	785	1,540	2,360	2,850
24	2	415	2,290	3,980	4,450	5,025
6	4	N.S.	755	1,095	1,225	1,720
12	4	15	955	1,675	2,325	2,600
24	4	400	2,420	3,980	4,550	4,540

Note: N.S. = Not Set
SOURCE: Dwight K. Smith, *Cementing* (Dallas: Spe-AIME, 1976)

Water ratio: 5.2 gal/sack
Slurry weight: 15.6 lb/gal

TABLE 5.2
EFFECT OF SODIUM CHLORIDE UPON THE THICKENING TIME AND COMPRESSIVE STRENGTH OF API CLASS A CEMENT

Water ratio: 5.2 gal/sack
Slurry weight: 15.6 lb/gal

	Thickening Time (hours: min.)			
	API Casing Cementing Tests for Simulated Well Depth (ft) of			
Sodium Chloride (percent)	1,000	2,000	4,000	6,000
0.0	4:40	4:12	2:30	2:25
2.0	3:05	2:27	1:52	1:13
4.0	3:05	2:35	1:35	1:20

		Compressive Strength (psi)			
		At Atmospheric Pressure, and Temperature of		At API Curing Pressure and Temperature of	
Curing Time (hours)	Sodium Chloride (percent)	40°F	80°F	800 psi 95°F	1,600 psi 110°F
12	0	70	405	1,065	1,525
24	0	940	1,930	2,710	3,680
48	0	2,110	3,920	4,820	5,280
12	2	290	960	1,590	2,600
24	2	1,230	2,260	3,200	3,420
48	2	3,540	3,250	3,900	4,350
12	4	280	1,145	1,530	2,575
24	4	1,390	2,330	3,150	3,400
48	4	3,325	3,500	3,825	4,125

SOURCE: Dwight K. Smith, *Cementing* (Dallas: SPE-AIME, 1976)

TABLE 5.3
COMMONLY USED CEMENT RETARDERS

Material	Usual Amount Used
Lignin retarders	0.1 to 1.0 percent*
Calcium lignosulfonate, organic acid	0.1 to 2.5 percent*
Carboxymethyl hydroxyethyl cellulose (CMHEC)	0.1 to 1.5 percent*
Saturated salt	14 to 16 lb per sack of cement
Borax	0.1 to 0.5 percent*

*Percent by weight of cement.

SOURCE: Dwight K. Smith, *Cementing* (Dallas: SPE-AIME, 1976)

because of low temperatures near the surface. At deeper depths and increased temperatures, the cement sets up too rapidly and the reverse problem is experienced. Most casing cementing jobs for deep wells require a cement slurry that can be pumped for about 2 to 4 hours. If the cement sets up before the job is finished, then cement will not be placed behind the pipe as desired. In order to measure slurry pumping time at higher temperatures, thickening time tests are run. The thickening time tests give an indication as to how long a slurry will be pumpable. If thickening tests indicate that a cement will not be pumpable long enough to complete a job, a retarder is added to increase the pumping time.

Materials commonly used as cement retarders are listed in table 5.3. Table 5.4 shows the retarding effect of calcium lignosulfonate on Class G or Class H cement. A review of this table indicates that the thickening time of Class G or H cement at 14,000 feet is 1 hour. This is not long enough for a normal casing job at this depth. By adding 0.4% calcium lignosulfonate to the cement, the thickening time is increased to about 3 hours. This will give adequate pumping time.

TABLE 5.4.
RETARDING EFFECT OF CALCIUM LIGNOSULFONATE ON API CLASS G OR H CEMENT SLURRIES

	Thickening Time (hours:min.) API Casing Cementing Tests for Simulated Well Depth (ft) of			
Retarder (percent)	8,000	10,000	12,000	14,000
0.0	1:56	1:26	1:09	1:00
0.2	2:15	2:12	1:38	1:25
0.3	3:38	2:40	2:14	1:58
0.4	4:42	3:36	3:10	2:58

SOURCE: Dwight K. Smith, *Cementing* (Dallas: SPE-AIME, 1976)

Lightweight Additives

Most cement slurries weigh about 15 to 17 pounds per gallon (lb/gal). Since there are many areas that can be drilled with 9.5- to 10.5-lb/gal muds, the hydrostatic head of the cement column will be much greater than the mud used to drill the well. The high hydrostatic pressure created by the cement may cause some weaker formations to break down and fracture. The slurry goes out into the fracture instead of filling up the space between the hole and the casing. The problem is especially bad on intermediate casing strings that are cemented all the way to the surface, as illustrated in the example problem in figure 5.1.

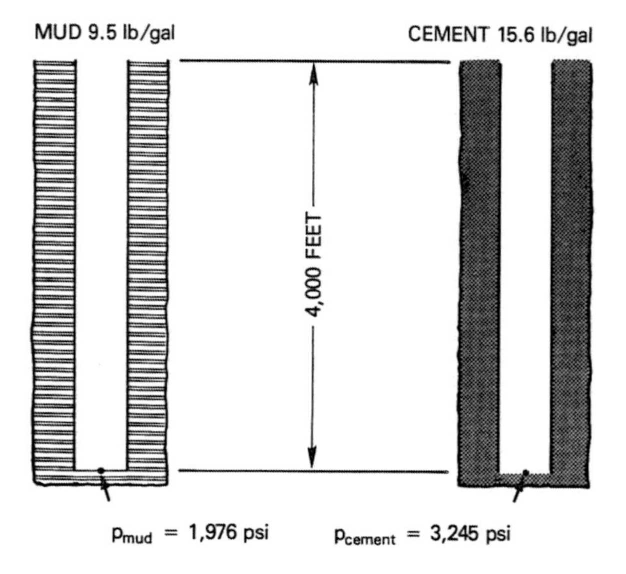

Figure 5.1. Pressure created by mud and cement columns

The pressure at the bottom of the hole is increased from 1,976 psi to 3,245 psi when 9.5-lb/gal mud is replaced in the annulus with 15.6-lb/gal cement. These pressures are calculated as follows:

Hydrostatic head (HH) = lb/gal × 0.052
 = psi/ft.

HH, mud = 9.5 × 0.052 = 0.494 psi/ft.

HH, mud = 0.494 psi/ft × 4,000 feet
 = 1,976 psi.

HH, cement = 15.6 × 0.052
 = 0.8112 psi/ft.

HH, cement = 0.8112 psi/ft × 4,000 feet
 = 3,245 psi.

Since the pressure at the bottom of the hole is increased from 1,976 to 3,245 psi, it is possible that this increased pressure might cause the formation to break down or fracture. Lightweight additives are added to the cement to decrease its weight and eliminate the problem of breakdown of weak formations. One common additive is bentonite. Bentonite can hold a large amount of water. It works in cement by tying up a lot of water that weighs only 8.33 lb/gal and thus causes a decrease of the cement slurry weights. Table 5.5 shows the effect of bentonite on Class H cement. Note that the slurry weight can be reduced from 15.6 lb/gal to 13.1 lb/gal by adding 8% bentonite. The slurry weight is reduced, but the compressive strength of the cement is also reduced. The 24-hour strength at 110° F of Class H cement without bentonite is 1,950 psi as compared to a strength of 255 psi with 8% bentonite under the same conditions. The reduced strength is acceptable under many conditions, and bentonite cements are used as "filler" cements quite often in high-column cement jobs.

Pozzolanic materials are also added to cement to decrease slurry weight. Natural pozzolans are of volcanic origin, and artificial pozzolans are manufactured from fly ash. Fly ash is a combustion by-product of coal. Sodium silicate, perlite,

TABLE 5.5
EFFECTS OF BENTONITE ON THE COMPOSITION AND PROPERTIES OF CLASS H CEMENT SLURRIES

Bentonite (percent)	Water Requirement (gal/sack)	Viscosity 0 to 20 minutes (Uc)*	Slurry Weight (lb/gal)	Slurry Volume (cu ft/sack)
0	5.2	4 to 12	15.6	1.18
2	6.5	10 to 20	14.7	1.36
4	7.8	11 to 21	14.1	1.55
6	9.1	13 to 24	13.5	1.73
8	10.4	12 to 19	13.1	1.92

Thickening Time (hours: min.)**

Bentonite (percent)	API Casing Cementing Tests for Simulated Well Depth (ft) of			API Squeeze Cementing Tests for Simulated Well Depth (ft) of		
	4,000	6,000	8,000	2,000	4,000	6,000
0	4:04	3:12	2:26	3:58	2:32	1:46
2	3:15	2:27	1:44	3:37	2:17	1:21
4	3:04	2:26	1:43	3:08	1:55	1:20
6	2:52	2:09	1:58	3:19	2:02	1:18
8	2:58	2:17	1:43	3:05	2:08	1:22

Compressive Strength (psi)

Bentonite (percent)	After 24 Hours at Temperature (°F) of				After 72 Hours at Temperature (°F) of			
	60	80	95	110	60	80	95	110
0	190	950	1,505	1,950	1,335	2,450	2,805	3,388
2	135	665	1,040	1,300	825	1,600	1,980	2,295
4	90	430	735	830	450	1,015	1,370	1,550
6	50	285	405	545	340	620	890	1,095
8	40	185	255	255	270	395	575	710

*Uc = units of consistency
**From pressure thickening-time tests

SOURCE: Dwight K. Smith, *Cementing* (Dallas: SPE-AIME, 1976)

and gilsonite are other materials used for the purpose.

Heavyweight Additives

Some wells require 16-lb/gal to 20-lb/gal muds for control during drilling. If this heavy mud is replaced by 15.6-lb/gal cement, loss of well control may result. In order to solve this problem, weighting materials such as hematite ore, barite, and sand are added to give additional weight to the cement slurry.

Lost Circulation Control

Lost circulation problems can occur with cements in the same manner in which they occur with drilling mud. The same solution is usually used, and lost circulation materials such as gilsonite, walnut hulls, or cellophane flakes are added.

Filtration Control

When a well is drilled with mud, some water (called filtrate) leaves the mud and goes into the formation. The mud that is left behind forms a *filter cake* and helps prevent more water from leaving the mud. The hole is eventually plastered over with the filter cake so that the hole will stand up and additional filtrate water will not be lost into the formation. The tendency of a mud to lose water is called *fluid loss*. Fluid loss is usually determined by seeing how much water will leak out of the mud through a filter in a test cell in a specified time at a specified pressure. Fluid loss is usually expressed as cubic centimetres/30 minutes (cc/30 min).

The same thing happens with cement except that it adds a new dimension to the problem of fluid loss. If water leaves the cement slurry, the cement will *flash set*. If the cement sets up, it can no longer be pumped, and the job must be aborted.

Another problem caused by the loss of filtrate water to the formation is formation damage. Since the filtrate water is usually fresh, the addition of the fresh water to the formation may cause some clays to swell, decreasing the permeability near the wellbore. Even if a saltwater cement is used, the saltwater filtrate can flow into the pore space and increase the water saturation. Increasing the water saturation will lower the oil permeability. This situation is called a *water block*.

Water loss in cement is usually lowered through the addition of an organic polymer.

Salt Cements

It has been mentioned that a small amount of sodium chloride (NaCl) added to cement will act as an accelerator. If larger amounts of NaCl are added, the NaCl will act as a retarder (fig. 5.2). The addition of about 5% NaCl decreases the thickening time for cement. At concentrations above 5%, the NaCl increases the thickening time or acts as a retardant.

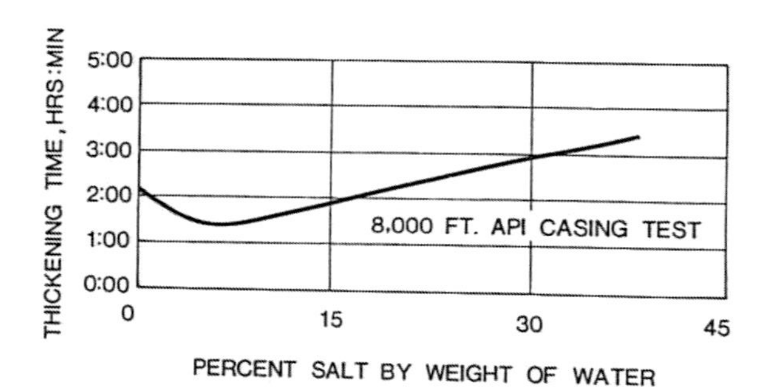

Figure 5.2. Effect of salt on thickening time and strength of API class G cement. From Dwight K. Smith, *Cementing* (Dallas: Society of Petroleum Engineers of AIME, 1976).

Salt-saturated cements are used to cement through the salt sections frequently encountered in drilling operations. Since the cement is already saturated with salt, the salty filtrate water won't wash out the salt sections.

Cements containing about 18% salt are commonly used, since the addition of the salt–

1. improves the bonding in shales;
2. acts as a dispersing agent; and
3. causes the cement to expand slightly when set, improving the bond to the formation.

Silica Flour

Cement loses its strength very rapidly at temperatures above 230° F. In order to prevent the loss of strength at high temperatures, silica flour is added to the slurry. Silica flour is a very fine sand, usually less than 200 mesh. The optimum amount of silica flour to add is 35% to 40%.

Factors That Affect Slurry Design

Up to this point only the properties of cements and the tailoring of a cement to fit the particular requirements of a well by the use of additives have been considered. In discussing additives, some of the factors that affect a slurry design were mentioned.

Pressure, Temperature, and Pumping Time

Increases in pressure and temperature decrease the pumping time of a cement slurry. Both pressure and temperature increase with depth. For example, on the Gulf Coast of the United States, the temperature increases about 1.9° to 2.2° F for every 100-foot increase in depth. Since the average ambient (surface) temperature is 74° F, a typical Texas or Louisiana Gulf Coast well will have a bottomhole temperature as follows, assuming a thermal gradient of 2° F/100 ft:

Depth, feet	*Temperature, °F*
5,000	174
10,000	274

$$\text{Temperature at 5,000 ft} = 74^\circ \text{ F} + \frac{5{,}000 \text{ ft}}{100 \text{ ft}} \times 2^\circ \text{ F} = 174^\circ \text{ F.}$$

The mud circulation will lower the temperature of the well until drilling is completed. The static temperature will warm back up to the original gradient after drilling is completed and mud circulation has ceased.

Cement Viscosity

The maximum strength for a cement slurry is achieved by mixing it with 2.8 gallons of water per sack. Unfortunately, cement mixed with 2.8 gal water/sack of slurry cannot be pumped. From the standpoint of placing the cement, a low-viscosity cement slurry is desired so that it can be pumped more easily. If the cement is mixed with too much water, free-water separation occurs, lowering strengths. Recommended water requirements for API cements and other cementing materials are given in table 5.6.

TABLE 5.6
WATER REQUIREMENTS

Material	Water Requirements
Class A & B Cement	5.2 gals. (0.70 cu. ft.)/94 lb. sack
Class C Cement (Hi Early)	6.3 gals. (0.84 cu. ft.)/94 lb. sack
Class D & E Cement (Retarded)	4.3 gals. (0.58 cu. ft.)/94 lb. sack
Class G Cement	5.0 gals. (0.67 cu. ft.)/94 lb. sack
Class H Cement	4.3-5.2 gals./94 lb. sack
Chem Comp Cement	6.3 gals. (0.84 cu. ft.)/94 lb. sack
Attapulgite	Approximately same as Bentonite
Ciment Fondu	4.5 gals. (0.60 cu. ft.)/94 lb. sack
Lumnite Cement	4.5 gals. (0.60 cu. ft.)/94 lb. sack
Halliburton Light Cement	7.7-10.9 gals./87 lb. sack
Trinity Lite-wate Cement	7.7 gals. (1.03 cu. ft.)/75 lb. sack (maximum)
Barite	2.64 gals. (0.35 cu. ft.)/100 lb. sack
Bentonite (gel)	1.3 gals. (0.174 cu. ft.)/2% in cement
Calcium Carbonate Powder	None
Calcium Chloride	None
Cal-Seal	4.8 gals. (0.64 cu. ft.)/100 lb. sack
CFR-1	None
CFR-2	None
D-AIR-1	None
D-AIR-2	None
Diacel A	None
Diacel D	3.3-7.4 gals./10% in cement (See Lt. Wt. Cement)
Diacel LWL	None (up to 0.7%) 0.8-1.0 gal./1% in cement (except gel or Diacel D slurries)
Econolite	Variable with concentration
GAS-CHEK®	None
Gilsonite	2.0 gals. (0.267 cu. ft.)/50 lb. cu. ft.
HALAD®-9	None (up to 0.5%) 0.4-0.5 gal./sk. of cement at over 0.5%
HALAD®-14	None
HALAD®-22A	None (up to 0.5%) 0.4-0.5 gal./sk. of cement at over 0.5%
Hi-Dense® No. 3	0.36 gal. (0.048 cu. ft.)/100 lb. sack or 3% by weight
HR-4	None
HR-5	None
HR-7	None
HR-12	None
HR-15	None
HR-20	None
Hydrated Lime	0.153 gal. (0.020 cu. ft.)/lb.
Hydromite	3.0 gals. (0.40 cu. ft.)/100 lb. sack
Iron Carbonate	None
LA-2 Latex	0-0.8 gals./sack of cement
NF-P	None
Perlite Regular	4.0 gals. (0.535 cu. ft.)/8 lb. cu. ft.
Perlite 6	6.0 gals. (0.80 cu. ft.)/38 lb. cu. ft.
Pozmix® A	3.6-3.9 gals. (0.48 cu. ft.)/74 lb. cu. ft.
Salt (NaCl)	None
Sand—Ottawa	None
Silica Flour (SSA-1)	1.6 gals. (0.21 cu. ft.)/35% in cement (32.9 lbs.) or 40% by weight SSA-1
Coarse Silica (SSA-2)	None
Spacer-Sperse™	None
Spacer Mix (Liquid)	None
SPHERELITE	95% by weight of SPHERELITE
THIX-SET Cement	5.2-13.8 gal./sack of cement
Tuf Additive No. 1	None
Tuf Additive No. 2	None
Tuf Plug	None

SOURCE: *Halliburton Cementing Tables*. Reprinted with permission of Halliburton Company.

Strength of Cement

The ultimate strength of a cement depends upon its composition. The strength also increases with age up to a point. Temperature also increases the buildup of strength to about 230° F, at which point the strength rapidly deteriorates. Increased pressure also increases the buildup of strength, but not to the same extent as temperature.

The strength of cement needed to support pipe is generally considered to be about 8-psi to 10-psi tensile strength. Since the tensile strength of cement is difficult to obtain by test, most tests are run on the compressive strength. A good rule of thumb is that the compressive strength of a cement is approximately eight to ten times its tensile strength. The minimum WOC time for cement can be estimated by seeing what time it takes to reach a compressive strength of about 80 to 100 psi.

Slurry Density

The need for density control was discussed with lightweight and heavyweight additives. Based on hole conditions, the density must be either increased or decreased to prevent problems. If well control is a concern, heavyweight additives may be added. If breaking down the formation is a concern, then the cement needs to be lightened.

Resistance to Downhole Brines

Many wells penetrate formations that contain sulfate-bearing water. Sulfate waters can cause strength loss and even complete failure of cement. In order to offset this problem, a sulfate-resistant cement should be specified. The sulfate resistance of a cement is governed by its composition. At the beginning of this chapter, a list of API cements is given, specifying the ones that are sulfate resistant.

Filtration Control

The cement filtrate water, not the cement itself, enters the formation. The loss of water to the formation can cause formation damage or cause premature setting of the cement. If water loss is expected to be a problem, then a suitable fluid-loss additive needs to be specified.

Cement Slurry Volume and Density Calculations

How Cement Slurries Are Specified

Cement volumes are specified by sacks. A sack of API cement weighs 94 lb and occupies a volume of 1 cubic foot; however, the actual volume occupied by the cement is only 0.478 cubic feet. The rest of the space, 0.522 cubic feet, is void space between the cement granules. Water fills in the 0.522 cubic feet of space between the granules.

Cement calculations are not difficult, but the way cement slurries are specified can be confusing. For example:

Slurry weight is expressed as lb/gal.

Slurry yield is expressed as cubic feet/sack.

Cement Slurry Yield and Density Calculations

Slurry density and yield calculations can best be illustrated by an example problem. Assume that a Class A cement is to be used in cementing the surface casing in a well. Table 5.7 shows that the weight of an API Class A cement is 94 lb/cu ft, and the absolute volume is 0.0382 gal/lb. The *absolute volume* is the actual volume occupied by the granules of cement, leaving out the pore space between the granules that will be occupied by water.

The absolute volume of water is 0.12 gal/lb. Since water weighs 8.33 lb/gal, absolute volume = 1/8.33 lb/gal = 0.12 gal/lb.

From table 5.6, it can be seen that a Class A cement requires 5.2 gal/sack.

Using the above data, table 5.8 can be prepared. The 94 lb of cement will occupy a volume of 3.6 gal.

TABLE 5.7
Physical Properties of Cementing Materials and Admixtures

Material	Bulk Weight (lbs/cu ft)	Specific Gravity	Weight 3.6* Absolute Gallons	Absolute Volume Gal/Lb	Absolute Volume Cu Ft/Lb
API Cements	94	3.14	94	0.0382	0.0051
Attapulgite	40.0	2.89	86.6	0.0415	0.0053
Ciment Fondu	90	3.23	97	0.0371	0.0050
Lumnite Cement	90	3.20	96	0.0375	0.0050
Trinity Lite-Wate	75	2.80	75.0‡	0.0429	0.0057
Barite	135	4.23	126.9	0.0284	0.0038
Bentonite (gel)	60	2.65	79.5	0.0453	0.0060
Calcium Carbonate Powder	22.3	2.71	80.9	0.0445	0.0059
Calcium Chloride, flake**	56.4	1.96	58.8	—	—
Calcium Chloride powder**	50.5	1.96	58.8	—	—
Cal-Seal	75	2.70	81.0	0.0444	0.0059
CFR-1**	40.3	1.63	48.9	—	—
CFR-2**	43.0	1.30	39.0	—	—
CFR-2 (Liquid)	—	1.18	—	—	—
D-AIR 1**	25.2	1.35	40.5	—	—
D-AIR 2**	—	1.005	—	—	—
Diacel A**	60.3	2.62	78.6	—	—
Diacel D	16.7	2.10	63.0	0.0572	0.0076
Diacel LWL**	29.0	1.36	40.8	—	—
Diesel Oil No. 1 (liquid)	51.1	0.82	24.7	0.1457	0.0195
Diesel Oil No. 2 (liquid)	53.0	0.85	25.5	0.1411	0.0188
Econolite**	75	2.4	72.1	—	—
Econolite (Liquid)	—	1.4	—	—	—
Flocele	15	1.42	—	—	—
GAS-CHEK® (Solid)**	72	2.7	80.9	—	—
GAS-CHEK® (Liquid)	—	1.07	—	—	—
Gilsonite	50	1.07	32	0.1122	0.0150
HALAD®-9**	37.2	1.22	36.6	—	—
HALAD®-14**	39.5	1.31	39.3	—	—
HALAD®-22A**	23.5	1.32	36	—	—
Hi-Dense® No. 3	187	5.02	150.5	0.0239	0.0032
HR-4**	35	1.56	46.8	—	—
HR-5**	38.4	1.41	41	—	—
HR-6L (Liquid)	—	1.21	—	—	—
HR-7**	30	1.30	39	—	—
HR-12**	23.2	1.22	36.6	—	—
HR-15**	44.4	1.57	47.1	—	—
HR-L (liquid)**	—	1.23	—	—	—
Hydrated Lime	31	2.20	66	0.0545	0.0073
Hydromite	68	2.15	64.5	0.0538	0.0072
Iron Carbonate	114.5	3.70	110.9	0.0324	0.0043
KCl (in solution @ 68°F. with fresh water)					
3%	—	1.019	—	0.0443	0.0059
5%	—	1.031	—	0.0450	0.0060
LA-2 Latex (liquid)	68.5	1.10	33	0.1087	0.0145
NF-1 (liquid)**	61.1	0.98	29.4	—	—
Perlite Regular	8***	2.20	66.0	0.0546	0.0073
Perlite Six	38†	—	—	0.0499	0.0067
Pozmix® A	74	2.46	74	0.0487	0.0065
Sea Water	—	1.025	—	—	—
Salt (dry NaCl)	71	2.17	65.1	0.0553	0.0074
Salt (in solution @ 68°F. with fresh water)					
6%—0.5 lb./gal.	—	—	—	0.0372	0.0050
12%—1.0 lb./gal.	—	—	—	0.0391	0.0052
18%—1.5 lb./gal.	—	—	—	0.0405	0.0054
24%—2.0 lb./gal.	—	—	—	0.0417	0.0056
Salt (in solution @ 140°F. with fresh water)					
Sat.—3.1 lb./gal.	—	—	—	0.0458	0.0061
Sand (Ottawa)	100	2.63	78.9	0.0456	0.0061
Silica Flour (SSA-1)	70	2.63	78.9	0.0456	0.0061
Spacer-Sperse™	40.0	1.32	39.6	—	—
Spacer Mix (Liquid)	—	.932	—	—	—
SPHERELITE††	25	.685	20.5	—	—
Coarse Silica (SSA-2)	100	2.63	78.9	0.0456	0.0061
Tuf Additive No. 1	—	1.23	36.9	0.0976	0.0130
Tuf Additive No. 2	15.38	.88	26.4	0.1364	0.0182
Tuf-Plug	48	1.28	38.4	0.0938	0.0125
THIX-SET A**	68.5	1.97	59.1	—	—
THIX-SET B**	36.5	1.37	41.1	—	—
Water	62.4	1.00	30.0	0.1200	0.0160

*Equivalent to one 94 lb sack of cement in volume.

**When less than 5% is used these chemicals may be omitted from calculations without significant error.

***For 8 lb of Perlite Regular use a volume of 1.43 gallons at 0 pressure.

†For 38 pounds of Perlite 6 use a volume of 2.89 gallons at 0 pressure.

††Varies with pressure.

‡75 lb = 3.22 absolute gallons.

SOURCE: *Halliburton Cementing Tables*. Reprinted with permission of Halliburton Company.

TABLE 5.8
Data for Calculating Density and Yield of Slurry

Material		Weight/lb	Absolute Volume (gal/lb)		Slurry Volume (gal)
Class A cement		94.0	0.0382		3.6
Water		43.3	0.12		5.2
	Total	137.3		Total	8.8

Cement volume = 94 lb × 0.032 gal/lb
= 3.6 gal.
Weight of water = 5.2 gal × 8.33 lb/gal
= 43.3 lb.
Volume of water = 43.3 lb × 0.12 gal/lb
= 5.2 gal.

From table 5.8, it can be seen that 8.8 gal of cement are the result of mixing 94 lb of cement and 43.3 lb of water. Since slurry density is specified as lb/gal,

$$\text{Slurry density} = \frac{137.3 \text{ lb}}{8.8 \text{ gal}} = 15.6 \text{ lb/gal.}$$

Slurry yield is specified in cubic feet/sack. Since there are 7.48 gal in a cubic foot,

$$\text{Slurry yield} = \frac{8.8 \text{ gal/sack}}{7.48 \text{ gal/cu ft}} = 1.18 \text{ cu ft/sack.}$$

The procedure outlined above can be used to calculate the density and the yield for cements containing other materials, such as bentonite and silica flour.

Cement Volume Calculations

After the correct cement and additives have been selected and the slurry yield and density calculations made, the next step in a casing cementing program is to determine the volume of cement slurry that will be required. This can best be illustrated by the following example problem:

Use Class A cement with 5.2 gal water/sack
Hole size = 8 3/4" (bit size)
Casing = 7" OD
Desired cement fill = 1,500 feet.

The theoretical volume between the hole and the casing can be calculated very easily on a hand calculator; however, it is much easier to use the tables that are in handbooks available from service companies.

From a service company handbook, the volume between the 7" OD casing and the 8 3/4" hole is found to be 0.1503 cu ft/linear ft.

Total annular volume = 1,500 ft
× 0.1503 cu ft/lin ft = 225 cu ft.

In the previous section, the yield of Class A cement with 5.2 gal water/sack was found to be 1.18 cu ft/sack.

$$\text{Amount of cement} = \frac{225 \text{ cu ft}}{1.18 \text{ cu ft/sack}} = 191 \text{ sacks.}$$

The above calculation of cement volume was based on the theoretical volume. In other words, it was assumed that the hole was 8 3/4", the same as the bit size. This matching of sizes is usually not the case, especially in soft formations. Some sections of the hole may be washed out, so it is necessary to determine the actual hole volume. This can be done by running a hole caliper.

Most open-hole electrical logs have a caliper that is run and recorded at the same time as the log. If a caliper log is available, it can be used to estimate the volume needed. However, two problems with using a caliper log are:

1. The caliper arms have a limited extension, so they will not record the deepest washouts.
2. It is difficult to look at several thousand feet of caliper log and come up with an average hole size.

The second problem can be solved by having the service company that ran the log integrate or calculate the average hole size. It will get an accurate average hole size measured by the caliper, but that still does not solve the problem of washouts larger than the caliper arm extension. Most engineers add an additional safety factor to account for washouts. The amount of safety factor added will vary greatly from area to area. A typical safety factor is in the range of 15% to 25%.

If a caliper is not available, then an additional amount of cement is added to the theoretical amount as a safety factor. The amount added depends upon the area, but typical safety factors are–

Type Hole	*Typical Safety Factor*
Surface casing	100%
Intermediate casing	50%–75%
Production casing	25%

In the problem, it was determined that 191 sacks were needed to fill up the 1,500 feet of annular volume between 7" casing and 8¾" hole. If a 50% safety factor were used, then it would be necessary to add an additional 96 sacks, making the total volume required 287 sacks. Since the calculations are not precise, a volume of 290–300 sacks should be used.

The above safety factors are listed to give an idea of how volume calculations are made. The actual safety factors used vary from field to field. They are also influenced greatly by the types of formations encountered. For example, unconsolidated sands have a greater tendency to wash out than hard, well-consolidated sandstone.

Placement Techniques

The next step in the casing cementing process is to pump the cement to the desired position in the hole. Casing cementing is basically a very simple process (fig. 5.3). The goal is to pump cement down the casing and displace all of the mud in the annulus to the desired cement height. A good bond should be obtained between cement and hole and between cement and pipe.

Although the above requirements for a good casing cementing job may sound relatively simple, it is difficult in actual practice to achieve all of the above goals. There are many problems encountered in the displacement process.

Displacement and Bonding Problems

One of the first problems encountered is due to the fact that mud and cement are non-Newtonian fluids. A Newtonian fluid, such as water, will start to flow as soon as pressure is applied, and its flow rate will be directly proportional to the applied pressure. A non-Newtonian fluid, such as drilling mud or cement, will not start to flow immediately when pressure is applied but will start in plug flow and progress to laminar and then turbulent flow as the flow rate is increased. Drilling mud, a non-Newtonian fluid, will "gel up" when it remains stationary, and the hole must be circulated to get the mud back into a liquid state.

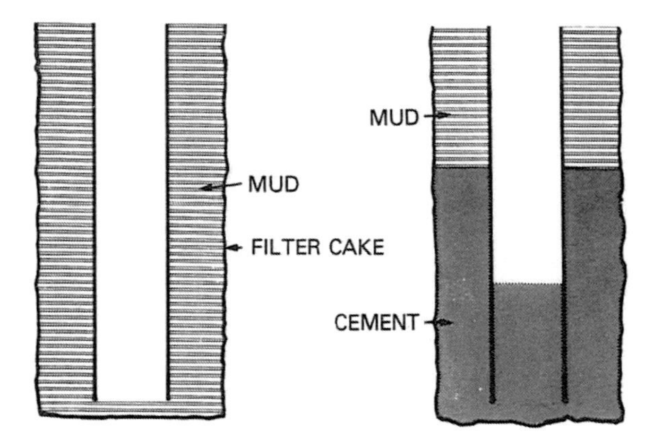

Figure 5.3. Casing cementing

Non-Newtonian fluids. The non-Newtonian fluids cause a problem illustrated in figure 5.4, which shows what can happen with off-centered pipe and non-Newtonian fluids. At low and moderate rates of flow, the mud, on the side where the standoff distance between pipe and hole is the least, does not flow. If mud is being displaced with cement and pipe is off-centered, the mud will not be displaced on one side of the pipe at low and moderate rates of flow.

Cement bond to hole. Another problem is the difficulty in obtaining a good cement bond to the

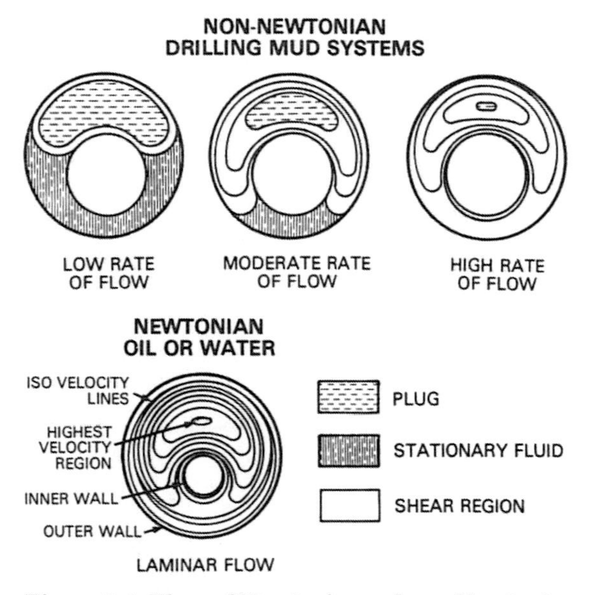

Figure 5.4. Flow of Newtonian and non-Newtonian systems in an eccentric annulus. From Piercy, N. A. V., M. S. Hooper, and H. F. Winney, "Viscous Flow through Pipes with Cores," *Phil. Mag.* 15, no. 99 (1933), 674.

formation because of the presence of the mud and the mud filter cake. The mud filter cake is a leatherlike material, which is plastered on the formation face in porous intervals. The filter cake prevents the cement from contacting the formation. Intimate contact between the cement and the formation is needed to get a good cement bond.

Cement bond to casing. A good cement bond to the casing is hampered by several factors. Casing comes from the mill with mill scale and varnish on its outer surface. These prevent good cement contact and make for a poor bond. The casing is also coated with mud, since the hole is full of mud when the casing is run. The presence of the mud coating also prevents cement contact with the pipe.

Solutions to Displacement and Bonding Problems

It is obvious that there are many obstacles to a good cement job. Fortunately, solutions have been found that solve or alleviate most of the problems.

Running clean casing. The cement bond to the pipe can be improved by removing the mill scale and varnish from the outside surface of the pipe. Mill scale and varnish are removed by chemicals or by sandblasting. A further improvement of the cement bond to the casing can be made by applying a resin-sand coating.

Conditioning mud. The mud displacement can be improved with conditioning the mud by circulating it to break any gels that have formed. This will minimize the possibility of mud contamination of the cement because of inefficient displacement of the mud by the cement.

Using a preflush. Preflushes are run ahead of the cement to clean the hole and remove loose wall cake. Preflushes are usually just water, although additives can be added to help thin the mud and penetrate the loose wall cake. A typical preflush is about 50 barrels of fresh water. A good rule of thumb is to use a preflush volume equal to 400 to 500 feet of annular fill. Cement itself is a good flushing agent, so the use of extra cement will also help clean the hole.

Running centralizers. The displacement problem due to off-centered pipe can be alleviated by running pipe centralizers. Centralizers are bow-spring devices run on the casing to keep it centered in the hole. Centralizers are normally run through and above and below the intervals of interest in the well. Charts are available to determine the distance between centralizers. A common spacing program is to run one centralizer for every joint of casing through the zones of interest. Caliper logs may be used to permit locating centralizers in zones that are full-gauge and not washed out.

Running scratchers. Scratchers are devices that attach to the casing and help remove the mud filter cake from the formation face. They are usually wire fingers or cables that are attached to the casing by metal bonds. By either rotating or reciprocating the casing, the wire fingers scratch the formation face and help remove the mud cake. By removing the mud filter cake, the cement bond to the formation is improved. Scratchers are run in porous zones where mud filter cake has built up on the formation face.

Utilizing pipe movement. Pipe movement is needed to operate the scratchers. Reciprocating-type scratchers are normally used, so the pipe movement is usually up and down. The up-and-down movement forces the wire scratcher fingers up and down on the mud filter cake, causing it to break loose. The reciprocating motion also helps in the mud displacement process.

Pumping in turbulent flow. Laboratory tests, as well as field experience, have shown that the displacement of the mud in the annulus by cement is enhanced by pumping in turbulent flow. All of the service companies have computer programs to calculate the pump rate needed to achieve turbulent flow. Pumping the cement in turbulent flow is an important factor in obtaining a good primary casing cementing job.

Allowing sufficient contact time. Laboratory tests have shown that a contact time of at least 10 minutes will improve the chances for a good cement job. Contact time is the time that cement flows past a point in the annulus. A contact time of at least 10 minutes ensures maximum mud removal. A sufficient amount of cement must be used to obtain this contact time over all of the zones of interest in the wellbore.

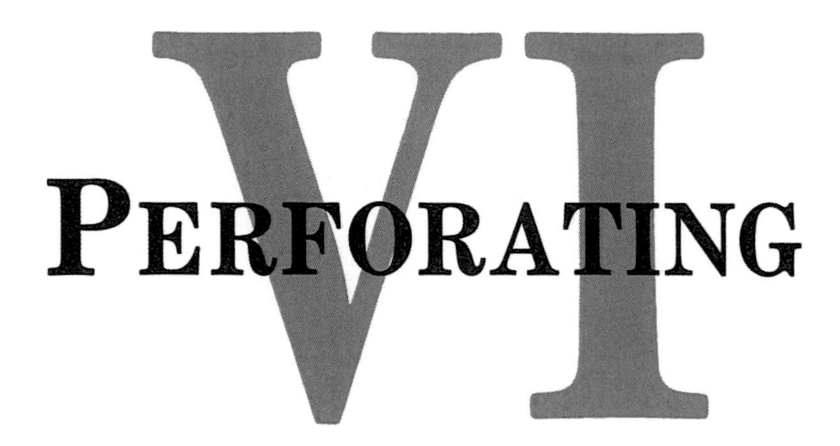

VI PERFORATING

The first well completions were of the open-hole type, and no perforating was required. The value of selective completions soon became apparent. The casing was set through the producing interval, and the producing interval perforated. The first perforating was performed with bullet guns using a steel projectile. With the development of the shaped charge for the World War II Bazooka, a new perforating technique became available, and the jet perforating system was developed. Bullet perforating is now very seldom used.

Jet Perforating Charges

Bullet perforating utilized a steel projectile that was fired from a gun similar to a rifle or pistol. The limitation was that the gun barrel had to fit in the casing of the well to be perforated. The jet perforator utilizes only a high explosive charge, and no projectile is used.

Principle of Operation

The explosive that is used in the jet perforator is a high-speed detonating type like TNT or nitroglycerin. The high explosive is detonated by a prima cord. The critical aspect of the shaped charge is the cone shape of the explosive. It is this cone shape together with the conical liner that is responsible for the remarkable perforating ability of this device.

The configuration of the explosives shown in *A, B* and *C* of figure 6.1 are similar except for the use of the conical hollowed-out sections in *B* and *C*. A steel target is used. Notice that a solid explosive charge has relatively little effect on the target (*A*). By hollowing out the center of the charge (*B*), a dramatic change takes place in the cavity created by the detonation of the explosive. However, it is still not suitable for oilwell use. The addition of the cone-shaped liner (*C*) is the key that makes the shaped charge into an effective tool for perforating steel, cement, and formation.

Figure 6.2 illustrates how a highly explosive jet stream forms from the detonation of the high explosive by the prima cord. It is this high-velocity jet stream that is capable of penetrating steel, concrete, and reservoir rock. The speed of the jet is 20,000 ft/sec and the pressure at the tip is 5 million psi.

Another important thing to observe in the figure is the formation of a slug, or "carrot," from the conical liner. The first jet perforators used a copper liner for the charge. A solid slug was formed by the disintegration of the copper liner, which partially plugged the perforation, resulting in less flow capacity from the perforation hole. In order to alleviate this problem, the liner is now made of powdered materials that do not form a solid slug and offer less restriction to flow from the perforation.

It is interesting to note that the slug forms in the back portion of the jet stream and has no effect on the perforation process. It only tends to clog up the hole that is formed by the jet. One misconception about jet perforating is that the high temperatures involved will fuse or glaze the reservoir rock. This is not correct; the

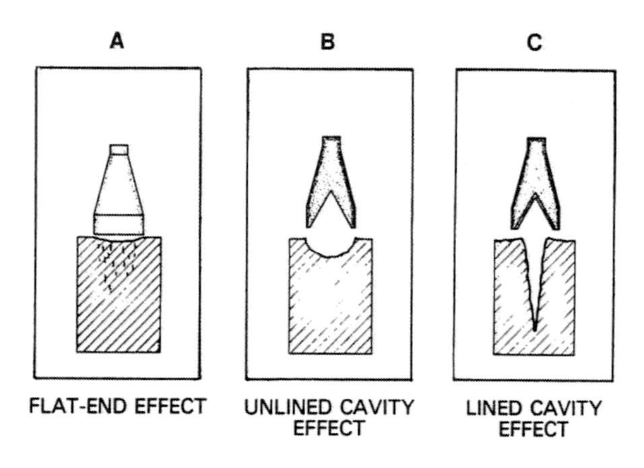

Figure 6.1. Unlined vs. lined cavity effects (Copyright 1972 by SPE-AIME)

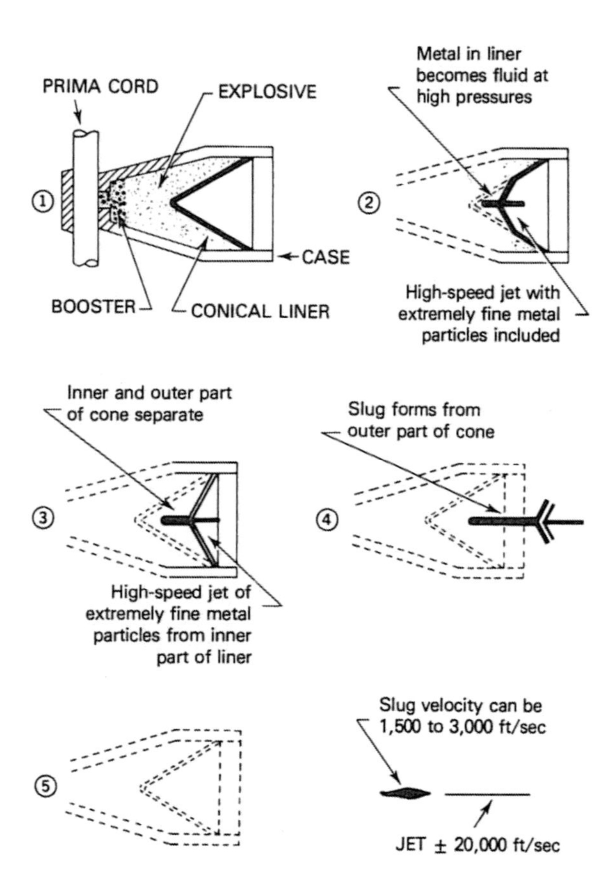

Figure 6.2. Jet perforating process using a solid metal liner. From Thomas O. Allen and Alan P. Roberts, *Production Operations: Well Completions, Workover, and Stimulation,* 2d ed. (Tulsa: Oil and Gas Consultants International, 1982)

temperatures involved are quite low. Jet perforators have been used to perforate wood or phone books to demonstrate that no burning occurs and that it is just the high-velocity jet stream that is responsible for the remarkable perforating ability of the jet perforator.

Effect of Charge Size on Penetration

The depth of penetration and the entrance hole size are generally proportional to charge size. A large charge will have a larger entrance hole and will penetrate deeper than a smaller charge of the same design. A small well-designed shaped charge may outperform a larger charge of inferior design, but generally the larger the charge, the better the performance.

Penetration Tests

One of the key factors in selecting a perforating charge is the depth of penetration that can be achieved. Originally each service company would conduct its own penetration tests by using targets it selected. Because of the lack of a standard target and uniform shooting conditions, it was difficult to compare the performance of charges of different companies. In order to correct this problem, the API adopted a standardized test procedure, API RP 43, to cover the testing of shaped charges.

API RP 43 specifies tests using two different targets. One target is Berea sandstone and another is a concrete, sand, and cement target. Since all service companies use the same test targets and the same shooting conditions for their tests, the tests can be used to compare the performance of different jet perforator guns.

Figure 6.3 shows the results of typical shaped-charge firing tests. It can be seen from a review of the test results that one portion of the form lists results of tests on the concrete target. Another portion of the form contains data concerning tests on the Berea sandstone target. To illustrate the use of the test results, consider the 2⅞" OD Hyper Dome charge. Refer to Section I for a concrete target; the average penetration is 16.12 inches and average casing hole (entrance hole) is 0.37 inches. Section II, which contains

CHARGE PERFORMANCE

Classification	Gun Size Inches	Gun Type	Charge Type	Rating psi/°F	Gun Phasing	Explosive Load (Grams)	Second Edition API RP-43 Sec I: Penetration (Inches) Sand-Cement	Casing Hole Diam.	Second Edition API RP-43 Sec II Berea Sand Formation (See Footnote): EH	OAP	TTP	TCP	CFE	ECP	Availability See Footnote
Thru-Tubing Expendable	1-11/16	Enerjet	Enerjet	20,000/ 340°	0°	10.00	11.67	0.310	0.34	7.62	6.76	5.63	0.83	4.75	3
	1-11/16	Uni-Jet	Aluminum (No Plug)	7,500/ 300°	0° /180°	10.5	6.36	0.33	0.34	6.23	6.23	5.11	0.88	4.49	1/4
	1-11/16	Uni-Jet	Ceramic (DPC)	15,000/ 300°	0° /180°	8.0	*5.11	0.34	0.30	4.80	4.80	3.68	0.83	3.07	1/4
	2-1/8	Enerjet	Enerjet	15,000/ 340°	0°	13.5	18.41	0.34	0.36	11.00	10.25	9.12	0.79	7.22	3/4
	2-1/8	Uni-Jet	Ceramic (DPC)	15,000/ 300°	0° /180°	16.2	*6.43	0.41	0.40	5.73	5.73	4.61	0.93	4.29	2
	2-1/8	Uni-Jet	Aluminum (Type A)	12,500/ 300°	0° /180°	13.5	15.93	0.31	0.33	11.03	10.17	9.01	0.73	6.63	3/4
	2-1/8	Uni-Jet	Ceramic (TypeA)	15,000/ 300°	0° /180°	15.5	*9.50	0.30	0.37	8.66	8.10	6.97	0.82	5.68	3
Thru-Tubing Hollow Carrier	1-5/16	Scallop	Hyper-Jet	15,000/ 340°	0°	1.7	3.95	0.22	0.23	2.93	2.62	1.50	0.70	1.06	3
	1-3/8	Domed Scallop	HyperDome	15,000/ 340°	0°	1.8	5.75	0.21	0.23	4.28	3.30	2.18	0.81	1.76	1
	1-11/16	Domed Scallop	HyperDome II	15,000/ 340°	0°	3.2	9.8	0.25	0.26	7.36	5.5	4.37	0.74	3.22	2
	2	Scallop	Hyper-Jet	15,000/ 340°	0° /180°	6.2	8.34	0.32	0.37	5.93	5.60	4.48	0.89	3.98	3
	2	Scallop	Hyper-Jet II	15,000/ 340°	0° /180°	6.5	9.45	0.33	0.37	7.02	6.50	5.38	0.81	4.36	3
	2-1/8	Domed Scallop	HyperDome	15,000/ 340°	0° /180°	6.5	10.64	0.32	0.34	8.83	7.92	6.80	0.76	5.13	1/4
	2-7/8	Domed Scallop	HyperDome	15,000/ 340°	0° /180°	14.0	16.12	0.37	0.40	10.98	10.63	9.50	0.76	7.44	1
Casing Expendable	3-1/8	Uni-Jet	Ceramic (DPC)	15,000/ 300°	0° /180°	20.5	*10.74	0.54	0.46	8.04	8.04	6.92	0.81	5.61	1
	3-1/8	Uni-Jet	Ceramic (Type A)	15,000/ 340°	0° /180°	20.5	14.69	0.42	0.42	10.47	9.85	8.72		6.89	3
	3-1/8	Uni-Jet	Aluminum (No Plug)	10,000/ 300°	0° /180°	27.0	10.31	0.62	0.55	8.54	8.54	7.41	0.82	6.08	1
Casing Hollow Carrier	3-3/8	Carrier	No Plug	20,000/ 340°	90°	11.0	9.52	0.37	0.40	8.13	7.30	6.17	0.77	4.75	2
	3-3/8	Carrier	Hyper-Jet II	20,000/ 340°	90°	14.0	19.36	0.38	0.38	11.30	10.85	9.73	0.81	7.88	2
	4	Carrier	No Plug	20,000/ 340°	90°	15.5	11.11	0.47	0.50	9.50	9.50	8.38	0.78	6.54	1/4
	4	Carrier	Hyper-Jet II	20,000/ 340°	90°	22.0	23.07	0.40	0.46	16.13	14.13	13.01	0.84	10.88	2
	4	Side Loaded Carrier	Hyper-Jet (Type A)	20,000/ 340°	90°	17.0	21.10	0.44	0.42	14.02	11.05	9.92	0.79	7.80	3/4
Selective Shot-by-Shot	3-3/8	Selective Casing Gun	Hyper-Select	20,000/ 340°	0°	10.0	17.70	.37	.37	10.8	9.5	8.38	.73	6.11	1/4
Sand Control Big Hole Gun	4	Carrier	Big Hole Hyper-Jet	20,000/ 340°	0°	22.0	4.32	0.89	0.91	4.60	4.48	3.36	0.73	2.47	1/4

CHARGE PERFORMANCE — HIGH TEMPERATURE PACKAGE

Classification	Gun Size Inches	Gun Type	Charge Type	Rating psi/°F	Gun Phasing	Explosive Load (Grams)	Penetration (Inches) Sand-Cement	Casing Hole Diam.	EH	OAP	TTP	TCP	CFE	ECP	Availability
Thru-Tubing Hollow Carrier	1-3/8	Domed Scallop	HyperDome	25,000/ 470°	0°	1.0	5.10		0.19	0.19	3.79	3.32	2.2	.64	2
	1-11/16	Domed Scallop	HyperDome	25,000/ 470°	0°	3.2	6.02	0.23	0.25	4.03	3.85	2.73	0.76	2.07	2
	**1-9/16	Scallop	Hyper-Jet	25,000/ 600°	0°	3.4	3.75	0.18	0.20	3.00	2.00	0.88	0.89	0.78	1
	2	Scallop	Hyper-Jet	25,000/ 470°	0° /180°	6.2	8.65	0.31	0.34	6.03	5.70	4.58	0.76	3.48	2
	2-1/8	Domed Scallop	HyperDome	25,000/ 470°	0° /180°	6.5	10.36	0.30	0.33	7.40	6.60	5.40	0.70	3.82	1
Casing Hollow Carrier	3-3/8	Carrier	Hyper-Jet	25,000/ 470°	90°	13.5	10.65	0.34	0.43	7.73	5.67	4.55	0.67	3.11	2
	4	Carrier	Hyper-Jet II	25,000/ 470°	90°	22.0	20.28	0.38	0.40	11.88	11.80	10.67	0.84	8.92	1

*1st Edition Data.
**Not officially published data.
API RD-43D certification-data sheets are available on request.

AVAILABILITY: 1 – Western Hemisphere
2 – Worldwide
3 – Eastern Hemisphere
4 – South America

EH = Entrance Hole
OAP = Overall Penetration
TTP = Total Target Penetration
TCP = Total Core Penetration
CFE = Core Flow Efficiency
ECP = Effective Core Penetration

Figure 6.3. Charge performance summary (Courtesy of Schlumberger)

data for the Berea sandstone target, shows that the average penetration (TTP) is 10.63 inches and average entrance hole diameter is 0.40 inches.

The test results contain some terms that require definition:

TTP: total target penetration as measured in inches.

TCP: total core penetration in inches. This number is equal to the TTP – 1⅛", since there is 1⅛" of steel and cement to penetrate before the core.

CFE: core flow efficiency. This reflects the ability of the perforation to transmit fluid, as compared with an ideal (drilled) perforation of the same diameter.

ECP: effective core penetration. This is $TCP \times CFE = ECP$. This number can be used to compare different charges, since it reflects not only the depth of penetration but also the ability of the perforation to transmit fluid.

Penetration of a jet perforator varies with the compressive strength of the reservoir rock. Berea sandstone was chosen for the tests because it is a uniform sandstone with little variation in its flow properties or compressive strength. The Berea sandstone has a compressive strength of 6,500 psi. If the formation to be perforated has a different compressive strength, the chart in figure 6.4 can be used to predict the estimated penetration.

The following example illustrates the use of the chart to estimate penetration in a reservoir rock.

Charge = 2⅞" Hyper Dome. See figure 6.3 for Berea sandstone target.

Average TTP = 10.63"

Formation = limestone with compressive strength of 13,000 psi.

A penetration of 8.5" is obtained at a compressive strength of 6,500 psi for the Berea sandstone and the midpoint of the 4" jet guns curve (fig. 6.4). The TTP for the Hyper Dome charge selected is 10.63", or 2.13" above this point on the chart.

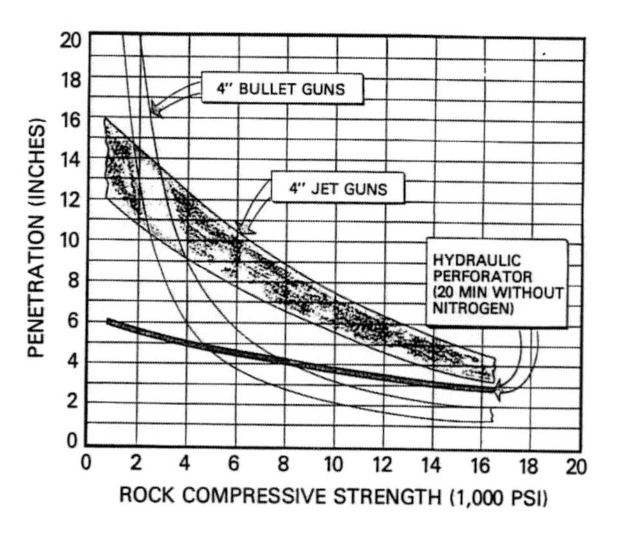

Figure 6.4. Relative performances of various devices (Copyright 1972 by SPE-AIME)

At a compressive strength of 13,000 psi for the limestone, a penetration of 5.0" is read at the midpoint of the jet guns curve. Since the Hyper Dome charge has a TTP = 2.13" above the midpoint of the jet guns curve, the estimated TTP is 5.0" + 2.13" = 7.13".

The TTP includes 1⅛" of steel and cement in the target. In order to determine the actual depth to which the formation is penetrated, the 1⅛" must be subtracted.

Formation penetrated = 7.13" – 1.125" = 6.00".

Types of Perforating Guns

There are two basic types of perforating guns:

1. Wireline guns, which are run in the hole on a braided wireline containing an electrical conductor
2. Tubing-conveyed guns, which are run in the hole on tubing.

Wireline guns are the ones most commonly used. They are cheaper, and the perforating interval can be more precisely located with the use of a radioactive log and a collar locator.

Tubing-conveyed guns are used in highly deviated holes in which wireline guns will not fall. They can also be used in those situations in which maximum gun size is needed and pulling the tubing after perforating is not desirable.

The discussion that follows is limited to wireline guns, but the principles governing jet perforating are the same regardless of the type of gun used. If more information on tubing-conveyed guns is desired, service company literature may be consulted.

Wireline perforating guns are of two types:

1. Casing guns: 3⅛" OD to 5" OD normal size
2. Tubing guns: 1$^{5}/_{16}$" OD to 2⅛" OD normal size.

The principal difference between the two types of guns is size. The gun that is run through tubing has to be smaller than the one run through casing. The above classification is based on normal tubing sizes up to 2⅞" OD. If larger-diameter tubing is used, then the classifications merge.

The tubing gun was developed so that wells could be completed by running the tubing, possibly with a packer, and installing the Christmas tree before the well is completed. It allows complete control over the well to be maintained.

If the well is perforated with a casing gun, it must normally be killed to run the tubing and packer and install the Christmas tree in the conventional manner. The tubing can be snubbed in under pressure, or a permanent packer with a blanking plug can be run and set above the perforated interval to permit the running of the tubing in the conventional manner. All of these methods that utilize casing perforating guns incur more risk than the use of a tubing gun with the tubing and Christmas tree in place.

Three types of guns in both casing and tubing sizes are available:

1. Hollow steel carrier
2. Fully expendable
3. Semiexpendable

Hollow Steel Carrier Guns

The hollow steel carrier gun is simply a steel tube fitted to hold jet charges. Its advantages are many.

1. It absorbs the detonation shock and thus minimizes casing damage.
2. It can be sealed and made gas-tight.
3. Its charge, blasting cap, and prima cord are protected from mechanical damage, so it is very reliable.
4. It has good pressure and temperature resistance.
5. It leaves no debris in the hole.
6. It is mechanically strong and can be spudded without damage.
7. It can be easily decentralized by magnetic or mechanical means.

Hollow steel carrier guns are used in the majority of perforating operations throughout the world.

Fully Expendable Guns

The fully expendable gun is usually made of aluminum-covered charges linked together. The principal advantage of this type of gun is its flexibility and its ability to shoot up to 200 feet on one run into the well.

One of the biggest disadvantages of the fully expendable gun is the standoff problem discussed in a subsequent section. This gun has a great potential for casing damage. The components of the gun are exposed to well fluids, increasing their susceptibility to failure. The aluminum cases are not gas-tight, and gas leakage can occur, greatly reducing the penetration ability of the charge.

A larger charge can be used with a fully expendable gun than with a hollow steel carrier gun of the same size. This advantage is offset to a great extent by the inability to decentralize the flexible expendable gun, resulting in a severe standoff problem.

Generally the use of expendable guns is limited to wells of shallow to moderate depth because of their temperature and pressure limitations.

Semiexpendable Guns

The semiexpendable gun consists of a row of charges held in a metal strip or between two stiff wires as carriers. The charges are usually encased in glass or ceramic coverings to protect them from downhole fluids and make them gas-tight.

Semiexpendable guns have several advantages over the expendable type. They are mechanically more substantial, and they leave less debris in the hole. The debris left is largely glass and ceramic, which do not cause as many problems as the aluminum left behind by expendable guns. However, the semiexpendable guns still present a serious potential for casing damage, similar to the expendable type.

Gun Selection

The hollow steel carrier gun is superior in many respects to the expendable or semiexpendable types. This fact accounts for its use on the vast majority of perforating jobs. Gun selection should be made on an individual well basis, and sometimes certain hole conditions dictate the use of the expendable or semiexpendable guns. For example, if there is a dogleg in the tubing string and it is impossible to get down with a hollow steel carrier gun, an expendable link-type gun is a good choice, since it is more flexible. The fully expendable or semiexpendable also has an advantage if a long section must be perforated on one run. But there is a greater risk of casing damage with the expendable or semiexpendable gun, and this must be a factor in the gun selection.

Factors Affecting Jet Perforating Performance

A number of factors besides those mentioned affect jet perforating performance. These need to be considered if optimum results are to be obtained.

Size and Design of Charges

Penetration is generally proportional to charge size. For charges of similar design, the larger charge will have the deepest penetration. If deep penetration is desired, then the largest charge designed for deep penetration should be selected. Some charges are designed to give maximum entrance hole diameter and not deep penetration.

Standoff Problem

API tests are conducted with the charge ½ inch from the target. If a small-diameter gun is used in casing, a standoff problem (fig. 6.5) can occur. The standoff problem normally occurs when through-tubing guns are used. A through-tubing gun is shown with each shot rotated 90° from the preceding shot. This rotation results in a spiral pattern, which repeats every four shots.

Since all holes are deviated to some extent, the tubing gun will lie on the low side of the hole, resulting in an effective shot pattern of two shots instead of four per foot (fig. 6.5, *left*). In other words, one-half of the shots are wasted. If all of the charges are pointed in the same direction (0° phasing), then all or none of the shots will be effective, depending upon which way the gun is pointing.

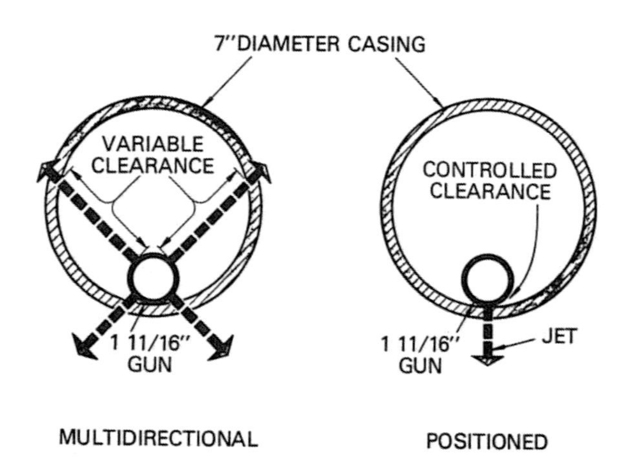

Figure 6.5. Standoff problem (Copyright 1972 by SPE-AIME)

This problem has been solved by using 0° phasing and then mechanically or magnetically decentralizing the gun so that all of the shots are directed toward the casing. The gun is positioned (fig. 6.5, *right*) so that each charge is held directly against the casing. The hollow steel carrier tubing gun is the most suitable one for decentralized shooting, since the hollow steel carrier will absorb the blast from the shaped charges and help prevent casing damage.

Hole Conditions

Equally as important as the charge and gun is the type of fluid in the hole during perforating and the direction of the differential pressure. If the pressure exerted by the hydrostatic pressure of the fluid in the hole is greater than the formation pressure, *overbalanced* conditions exist. If the pressure in the hole is less than the formation pressure, then *underbalanced* conditions are present.

Effect of well fluids. The worst type of perforating fluid is mud. Usually the hydrostatic pressure of the mud column is greater than the formation pressure. When the well is perforated, the excess pressure inside the casing drives the mud into the newly created perforations. Since drilling mud is designed to plug, plugging is precisely what happens. The drilling mud fines are forced into the perforation and cause a blockage problem. This problem can be solved by using a clean fluid such as filtered salt water as perforating fluid. If salt water compatible with the formation is pushed into the perforation, it will not have any solids to plug the perforations.

For perforating in a limestone reservoir, acetic or hydrochloric acid can be used as the perforating fluid. When these acids are pushed into the perforations, they react with the rock debris and assist in the perforation cleanup.

Effect of pressure differential. Perforating in mud results in mud particles being forced into the perforations, causing plugging and reduced flow performance. Even if clean salt water is used as a perforating fluid, damage can occur. Apparently when the formation is perforated, rock debris is formed by the jet. In addition, the rock around the jet stream is compressed. If the pressure is reduced by an immediate flow from the formation to the wellbore, the formation will tend to "spring back" and assist in the cleanup. If the pressure differential is from the wellbore toward the formation, the flow will be from wellbore to formation after perforating. This will keep the compressed formation from "springing back" and will also drive the debris into the pore openings, reducing flow capacity.

It is obvious then that the best perforating efficiency will be achieved when the pressure differential is from the formation to the wellbore. Oilwells are normally shot with a 200-to 500-psi pressure differential toward the wellbore. Some low-permeability gas wells have been shot with differential pressures as high as 2,000–3,000 psi into the wellbore.

The effects of pressure differential on core flow efficiency is shown in figure 6.6. Note that a core flow efficiency (CFE) of a little over 0.7 is achieved with a 200-psi differential into the wellbore. If the differential is reduced to 25 psi, the CFE drops to about 0.3. Also note that the curve flattens out at 200 psi. It is for this reason that a differential of 200 to 500 psi toward the wellbore is used in many perforating jobs.

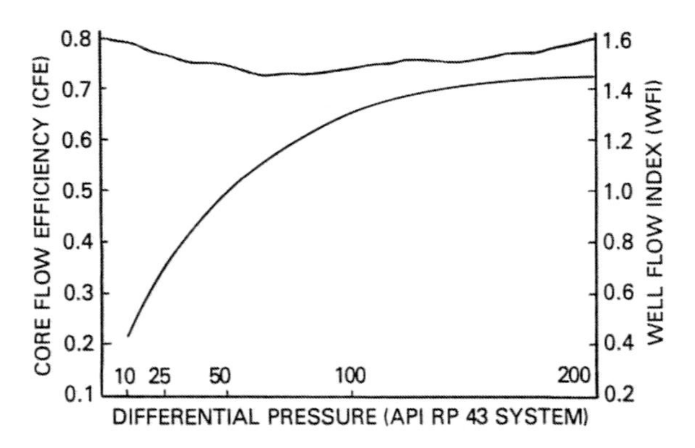

Figure 6.6. Effects of pressure differential on core flow efficiency (Copyright 1972 by SPE-AIME)

Penetration Depth and Shot Density

Up to now it has been implied that penetration depth is important, and generally the maximum penetration possible is desirable. This fact is true up to a point. The effect of shot density and penetration depth on productivity is illustrated in figure 6.7. The productivity ratio q_p/q_r is plotted vs. penetration depth. The ratio q_p/q_r indicates the flow capacity of the perforation divided by

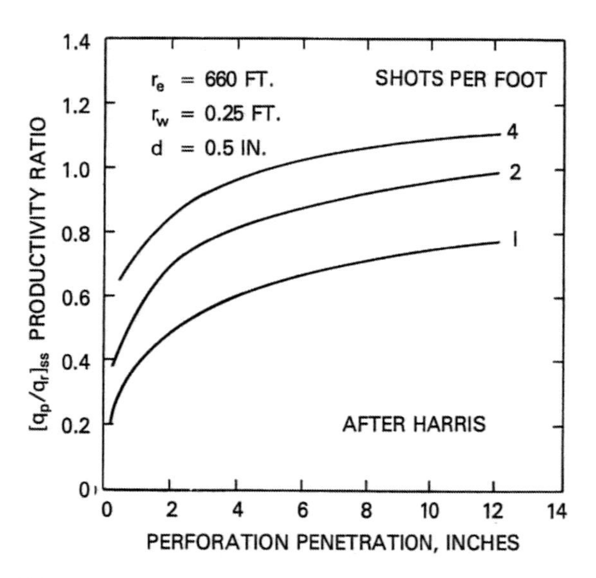

Figure 6.7. Effect of shot density and penetration depth on productivity (Copyright 1972 by SPE-AIME)

the flow capacity of a drilled or damage-free perforation.

One interesting aspect of the figure is that values of q_p/q_r greater than 1 are plotted. Values greater than 1 are possible because the perforation extending out into the formation increases the effective diameter of the wellbore. The discussion of Darcy's equation for radial flow showed that an increase in wellbore diameter will increase the flow rate.

Note that the curves flatten out after about 6 inches of penetration and that penetration deeper than 6 inches does not increase the productivity ratio appreciably.

Another thing to observe on this chart is that shot density is more important than penetration. Four to 5 inches of penetration with four shots to the foot gives about the same productivity ratio as 12 inches of penetration with two shots to the foot.

Figure 6.7 assumes an undamaged formation. If a severe formation damage problem is present, then deeper penetration is more important (fig. 6.8).

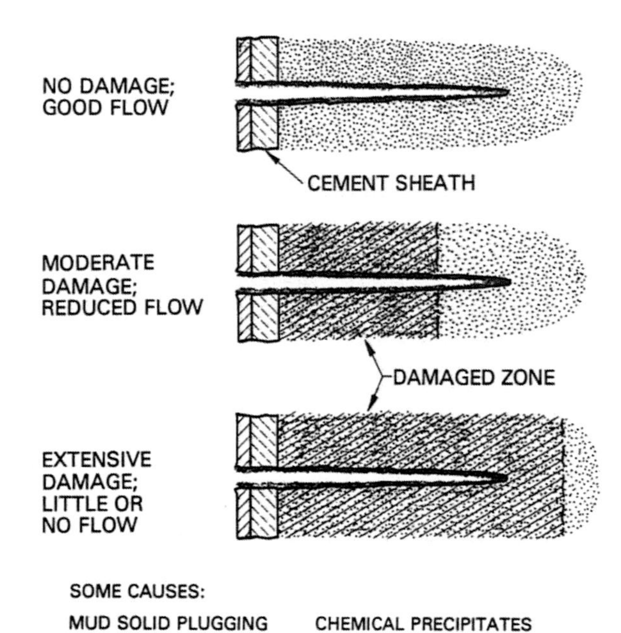

Figure 6.8. Wellbore damage or skin effect (Copyright 1972 by SPE-AIME)

Shot Orientation

Shot orientation is an important aspect of oilwell shooting and needs to be considered in more detail. Figure 6.9 illustrates the effect of shot orientation on perforation productivity. The orientation of the charges is referred to as *phasing*. For example, the $m = 1$ orientation is called *zero phasing*, since all shots are lined up on the same side. The orientation in which $m = 2$ is called *180° phasing*. One-half of the shots face one direction and the other half face the opposite direction. The orientation in which $m = 4$ is called *90° phasing*, since each shot is positioned 90° from the preceding shot.

Note that 0° phasing with four shots per foot will result in a productivity ratio of 0.98. This is essentially the equivalent of the productivity of an open-hole completion. Note also that there is very little difference between the productivity at 180° and 90° phasing with four shots per foot. Productivity ratios from 180° and 90° phasing are both equal to about 1.1.

The 90° phasing depicted in the diagram shows all four shots on the same plane. This type of configuration is not usually used in perforating for production. It is sometimes called a *four-way gun* and is used to perforate prior to a cement squeeze.

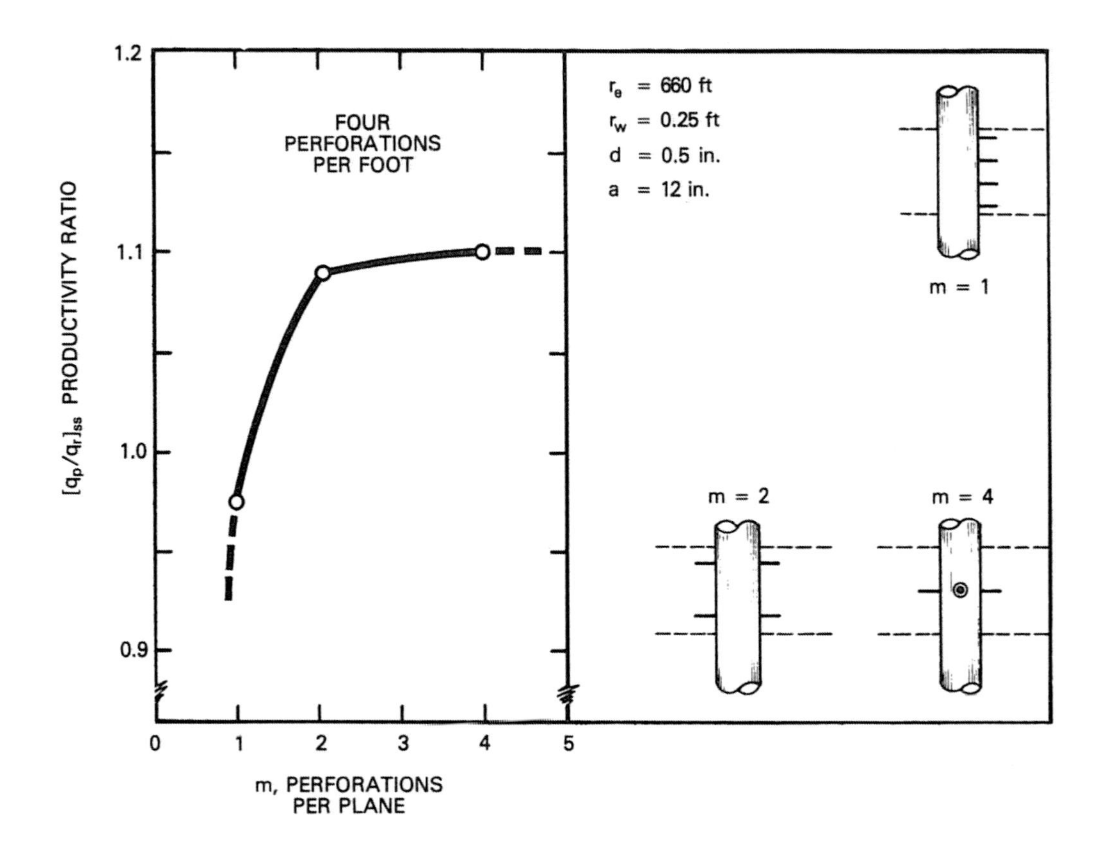

Figure 6.9. Effect of shot orientation on perforation productivity (Copyright 1972 by SPE-AIME)

The 90° phasing used to perforate for production has the perforations in a spiral pattern, 90° apart. With four shots to the foot and 90° phasing, each shot will be displaced 3 inches and 90° from the previous shot. The productivity ratio will be the same for the spiral pattern as for the four shots in one plane. If the formation is not homogeneous, the spiral pattern should increase the chances of getting the shots into good permeability.

Casing Damage Potential

An expendable charge in contact with the casing when it detonates can cause the casing to be split or bulged. The potential for splits and bulges can be eliminated by using a hollow steel carrier gun. If using the steel carrier is not feasible, the charge should be as small as possible, since the amount of damage will normally be less with the smaller charges.

Optimum Conditions for Perforating

Now that the various factors affecting jet perforating efficiency have been considered, it is possible to prepare the following list of optimum jet perforating conditions.

1. *Charge:* Using the largest possible well-designed charge that will give maximum penetration.
2. *Gun:* Using a hollow steel carrier gun to minimize casing damage.
3. *Perforating fluid:* Perforating in a clean, nondamaging fluid.
4. *Pressure differential:* Perforating with a pressure differential into the wellbore. Pressure differentials of 200 to 500 psi are commonly used. With proper precautions, pressure differentials as high as 2,500–3,500 psi have been used in tight gas sands.

5. *Shot density:* Shooting with four shots per foot for maximum productivity.
6. *Shot pattern:* Shooting with 90° phasing.
7. *Standoff distance:* Perforating with the standoff distance as small as possible.

It is obvious from a review of the above optimum conditions that the only way to achieve all seven is to use a hollow steel carrier casing gun. There are many instances when the use of a casing gun is not desirable or feasible and a compromise has to be made. In fact, most perforating jobs are compromises, and only the most important of the optimum conditions are observed. The important thing is to be aware of all seven.

Perforating Program Selection

The selection of a perforating program can best be illustrated by two examples, using two widely different conditions.

Perforating High-Permeability Formations

Perforating jobs for wells completed in high-permeability sands are usually not too critical. Formation damage is seldom a problem, and neither is well productivity. In order to illustrate a typical job, assume that a well with the following conditions is to be perforated.

Well depth	10,000 ft
Bottomhole pressure	4,700 psi
Expected shut-in surface pressure	3,800 psi
Average permeability	275 md
Average depth of formation damage	2 in.
Pay thickness	50 ft
Formation	Poorly consolidated sand
Type of production expected	Gas condensate

Area experience: Previous wells completed in this formation have had no formation damage when killed with salt water to run tubing after perforating. Wells are produced through tubing set on a packer.

General: The economics of development are not good because of limited reserves. It is desired to minimize completion costs as much as possible consistent with safe operation practices.

1. and 2. *Charge Size and Gun Type*

It is necessary to consider charge size and type of gun at the same time because their selections are interrelated. A through-tubing type of perforating job is indicated by the data. The well has a fairly high expected shut-in surface pressure, and it would be desirable to perforate with the tubing and Christmas tree in place. Another factor suggesting a tubing gun is economics. Completion costs can be decreased by running the tubing and packer, installing the Christmas tree, and then releasing the drilling rig. The well can be completed with a small mast for the wireline unit.

A tubing charge that has more than 2 inches of formation penetration should be selected to penetrate the damaged zone near the wellbore.

2. *Perforating Fluid*

Clean salt water should be used. Since the well will be shot with a pressure differential into the wellbore, this fluid should be satisfactory.

3. *Pressure Differential*

A pressure differential to the wellbore of 200 to 500 psi should be used. This will give maximum productivity.

4. *Shot Density*

Four shots per foot should be used unless there is some reason to use a lesser density.

5. *Shot Pattern*

A 0° phasing should be used. Hollow steel carrier tubing guns should always be shot with 0° phasing if they are decentralized.

6. *Standoff Distance*

The gun should be decentralized in order to have zero standoff distance. Hollow steel carrier guns with 0° phasing should always be decentralized.

Note from reviewing the above recommendations and the list of optimum perforating practices that five of the seven optimum conditions have been realized. It is not necessary to achieve all seven optimum conditions except on occasional jobs. Be aware of all seven and pick the necessary ones based on the individual well conditions.

Perforating Low-Permeability Formations

Generally, low-permeability reservoirs are stimulated by acidizing or fracing. In this case the type of stimulation may dictate what type of perforating is selected. If the well is stimulated by using the limited-entry technique, as many holes as needed for the treating rate contemplated are perforated.

In many instances it is desirable to obtain a natural test of a well before stimulation. In this type of situation a deeply penetrating charge is desirable. Filtrate waters from drilling muds and cement tend to invade low-permeability rock deeper than they do high-permeability rock. In view of this, wellbore damage generally extends deeper into the formation in the low-permeability rock. A casing gun with the largest and deepest penetrating charge available should be used to perforate low-permeability zones to be tested naturally. Since a natural test at the highest rate possible is desired, it is usually best to strive for all seven optimum conditions.

The following example illustrates proper selection of perforating conditions for a low-permeability sand when a natural production test is desired.

Well depth	10,000 ft
Bottomhole pressure	4,700 psi
Shut-in surface pressure	3,800 psi
Average permeability	0.5 md
Average depth of formation damage	6 in.
Gross pay thickness	400 ft
Net pay	50 ft
Formation	Hard, dense, fine-grained sandstone
Casing size	5½" OD
Type of production expected	Dry gas

Area experience: Previously completed wells in this formation have experienced severe formation damage when they are killed to run tubing after perforating. Because of the expected high shut-in surface pressure, wells are completed with the tubing set on a packer to keep the high shut-in pressure off the casing.

General: Commercial wells can be made by fracing with large-volume, high injection-rate frac jobs. Because of the large expense of these jobs, a natural flow test and a shut-in pressure buildup test are run to evaluate the suitability of the well for fracing.

1. and 2. *Charge Size and Gun Type*

Since maximum penetration is desired in order to penetrate the damaged zone and achieve the highest productivity possible on the natural test, the largest charge designed for deep penetration should be used. This will require the use of a casing gun. A hollow steel carrier should be used to minimize casing damage and permit the use of 90° phasing for maximum productivity.

3. *Perforating Fluid*

A clean, filtered salt water should be used to prevent formation damage.

4. *Pressure Differential*

A high pressure differential to the wellbore, 2,500–3,500 psi, should be used. This will require special precautions during perforation, but no problem should be encountered.

5. *Shot Density*

Four shots per foot should be used for maximum productivity.

6. *Shot Pattern*

Shooting with 90° phasing should be done to get maximum productivity and uniform coverage of the heterogeneous formation.

7. Standoff Distance

The largest-diameter casing gun that can be run in the 5½" OD casing should be selected so that the standoff distance will be minimized.

The perforating program selected achieves all seven optimum perforating conditions. The use of the casing gun and underbalanced conditions will necessitate the use of a snubbing unit to run tubing under pressure for the natural flow test. As an alternative, a permanent packer with a blanking plug could be run and set above the perforated interval to allow the tubing to run in the normal way. The extra expense and additional problems can be justified, since a natural flow test and buildup will furnish data to evaluate an expensive frac job.

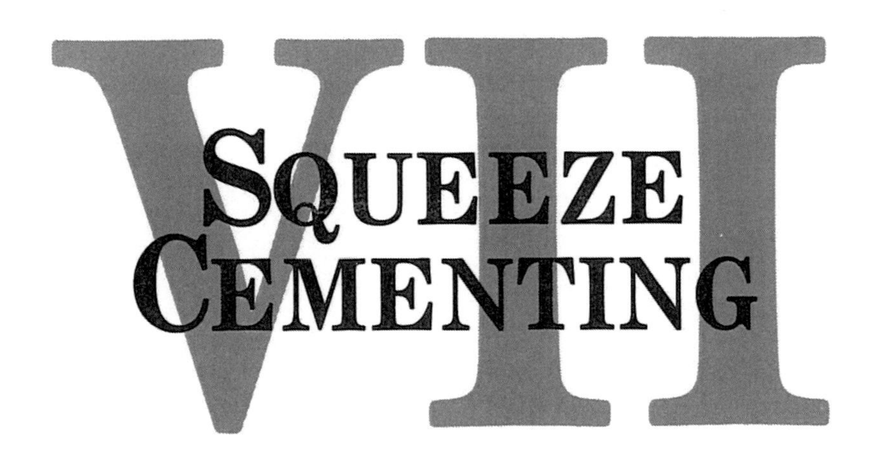

Squeeze cementing is a remedial procedure that involves pumping cement under pressure into a formation void or against a porous zone in order to make a seal between casing and formation. It is used to repair casing leaks, to control excessive gas or water production, to abandon a producing interval, and to repair primary cementing failures. Squeeze cementing techniques are of two types: (1) high-pressure squeeze and (2) low-pressure squeeze. The high-pressure squeeze was the method originally developed and the only one available until the early 1950s. The theory behind high-pressure squeeze jobs was based on a number of conceptions that have since been discredited.

High-Pressure Squeeze

High-pressure squeeze is invariably performed under a packer, since high pressures are involved (fig. 7.1). To illustrate the method, assume that the goal is to shut off bottom water in the well. The usual procedure is to squeeze off all of the perforations and then drill out and reperforate in the top of the sand to minimize water production.

After the packer is set, fluid is pumped down the tubing to break down the formation and establish a pump rate. In the past, mud was a popular breakdown fluid; however, water is more commonly used today. After the formation is broken down, usually between 100 and 200 sacks of cement are pumped. If a pressure buildup is encountered, the surface pressure is increased to some arbitrary value like 4,500 psi. The fact that a high surface pressure is attained is generally considered to be a sign that a successful job has been performed. The packer is then unseated, and the excess cement is reversed out.

If a surface pressure buildup is not encountered, the *hesitation* method is used. The hesitation process is accomplished by shutting down the pumps for a short interval, then starting to pump again. This process is repeated until pressure buildup is attained or the supply of cement is exhausted. If the latter occurs, more cement can be mixed, and the hesitation process can be continued.

Although no statistics are available regarding the success ratio of high-pressure squeezes, it is probably around 50%.

Development of Cement Squeeze Concepts

When high-pressure squeeze jobs were first used, no knowledge was available regarding hydraulic fracturing, and little thought was given to the path taken by the cement pumped through the perforations. With the advent of hydraulic fracturing in the early 1950s, the theory that cement was pumped into the formation in horizontal "pancakes" became popular. This theory supported the practice of breaking down the formation, since it was believed that the horizontal pancakes of cement would prevent vertical fluid migration. As fracturing

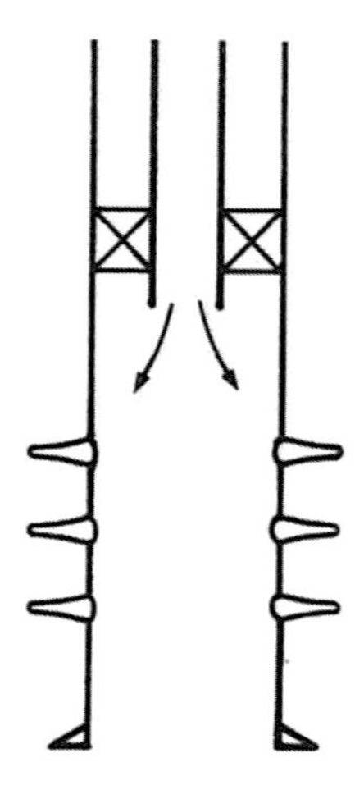

Figure 7.1. Typical high-pressure squeeze

technology improved, more evidence became available showing that hydraulically created fractures are vertical except in some instances in shallow wells. This development cast serious doubt on the desirability of high-pressure squeeze jobs. At about the same time, a new technological development ushered in the low-pressure squeeze technique.

Low-pressure squeezes were developed in the mid-1950s by Exxon (then Humble) as part of their permanent well-completion technique. The concept of the permanent well-completion system was to make an initial completion and set the tubing, allowing all subsequent work to be performed without removing the tubing string. As part of this system, Exxon developed a low-pressure squeeze technique using low-water-loss cement to squeeze off perforations. Cement was spotted across the perforations using a small-diameter string of tubing. While the cement was still in a fluid state, any excess was reversed out. Only "buttons" of cement were left in the perforations to make a seal. Although there were problems with this technique, it was generally successful, and the success caused many people to rethink high-pressure squeeze cementing.

When squeeze cementing was first used, a popular conception was that whole cement is pumped into the formation. However, formation rock is like a filter. It has small openings through which fluids flow, but solid materials like cement are screened out at the wellbore. Unless the formation is fractured, only cement filtrate water enters the formation. The only way that the cement can be made to go into the reservoir rock is to fracture the formation; then the cement can be pumped into the fractures.

The fact that only the filtrate water enters the formation is important. If ordinary cement is pumped into perforations at a pressure below the fracture pressure of the formation, the cement will displace any fluid that is in the perforation, and then the flow of cement will stop. If continued pressure is applied, the water will be squeezed out of the cement in the perforation, and filtrate water will enter the formation. When the water is removed from the cement, the cement will set up. And when part of a slurry sets up because of loss of water, the cement near it will "flash-set."

In a high-pressure squeeze job, if 100 to 200 sacks of cement are pumped through a set of perforations, the cement goes out into the formation in a fracture. When hydraulic fracturing technology made this fact apparent, high-pressure squeeze theory was adjusted, and it was assumed that flat pancakes of cement were pumped into the formation into fractures with a horizontal orientation and that horizontal pancakes of cement would prevent the vertical migration of fluids.

However, data began to be collected through the use of downhole cameras and inflatable packers to show that, except for relatively shallow wells above 2,000 feet, hydraulically induced fractures have an essentially vertical orientation. Instead of preventing vertical migration of fluids, high-pressure squeezes with horizontal, flat pancakes of cement tend to promote it.

Low-Pressure Squeeze

In performing a low-pressure squeeze, assume that the goal is to shut off bottom water by squeezing off the current perforations. After they are squeezed off, it is necessary to reperforate the upper part of the pay section to try to restore production with reduced water content.

A low-pressure squeeze job can be performed with or without a packer, since low pressures are involved. Assume that a packer is used (fig. 7.1). After the packer is set, cement is pumped down the tubing and into the perforations. The fluid in the tubing, the casing below the packer, and the perforations is displaced by the cement into the formation. The amount of fluid displaced into the formation can be minimized by not setting the packer until the cement is on bottom. As soon as the cement fills the perforations and any voids, the pressure will start to rise. Generally a few hundred psi of surface pressure is applied at the surface, being careful that the fracture pressure of the formation is not exceeded. The packer is then unseated.

At this point there are two alternatives. The cement can be left in place across the perforations and allowed to harden. At a later date it can be drilled out so that the top of the pay section can be reperforated. Another option is to reverse out the cement in the casing through the perforated interval while it is still in a fluid state. This will eliminate the necessity to drill out the cement to permit reperforating the upper part of the pay section. The perforations are shut off by the buttons of cement left in them.

The low-pressure squeeze, then, is basically a very simple process. It does have one very critical aspect, and that is the control of the water loss in the cement. Recall that when the cement enters a perforation and is subjected to pressure, cement filtrate water is squeezed out of the cement into the formation. This process causes the nearby cement to flash-set. If a squeeze job is performed with ordinary cement, flash-setting is exactly what happens. If the cement enters an upper perforation, the cement will set up in the casing and prevent cement from entering the lower perforations (fig. 7.2). Note that there is solid cement in the casing opposite the upper perforations, and mud remains in the casing opposite the lower perforations. Since there is solid cement in the casing at the upper perforations, pressure applied at the surface will not force cement into the lower perforations. This fact explains why 4,500 psi can be applied at the surface on a high-pressure squeeze job and a failure is still possible in drilling out and testing.

The answer to the problem illustrated in figure 7.2 is to use a low-water-loss (LWL)

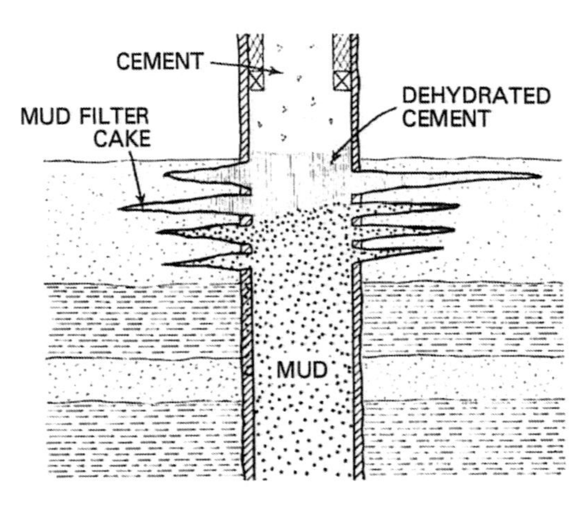

Figure 7.2. Slurry dehydration across open perforations during a high-pressure squeeze operation. From Dwight K. Smith, *Cementing* (Dallas: Society of Petroleum Engineers of AIME, 1976).

cement and apply a surface pressure that will not fracture the perforated formation. The need for water-loss control is illustrated in figure 7.3, which shows what happens when cements with different water losses are pumped into a perforation. In the top perforation, cement with a water loss of 1,000 cc (no water-loss control) is pumped. Note that the cement has set up in the casing, effectively blocking it off. A cement with a 300-cc water loss has been pumped into the

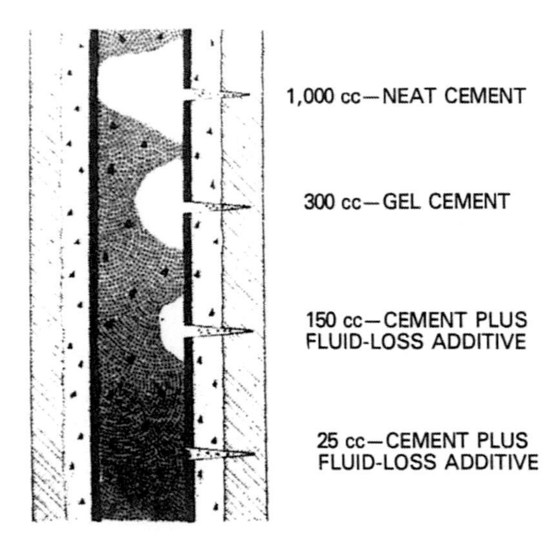

Figure 7.3. Cement node buildup effect with filtration control, showing API fluid loss in cc/30 minutes at 1,000 psi. From Dwight K. Smith, *Cementing* (Dallas: Society of Petroleum Engineers of AIME, 1976).

second perforation. A large globule of cement has set up in the casing, but the casing is not blocked. It can be seen that with a 25-cc water-loss cement, the cement has not set up in the casing, but only in the perforation. It is obvious that filtration control is the key to successful low-pressure squeeze jobs.

Note in figure 7.4 that nodes or buttons of cement have formed in the perforations, but the slurry in the casing is still fluid. The perforations are closed off, and yet the cement in the casing is still fluid. The cement inside the casing can be circulated out of the hole, eliminating the necessity to drill out.

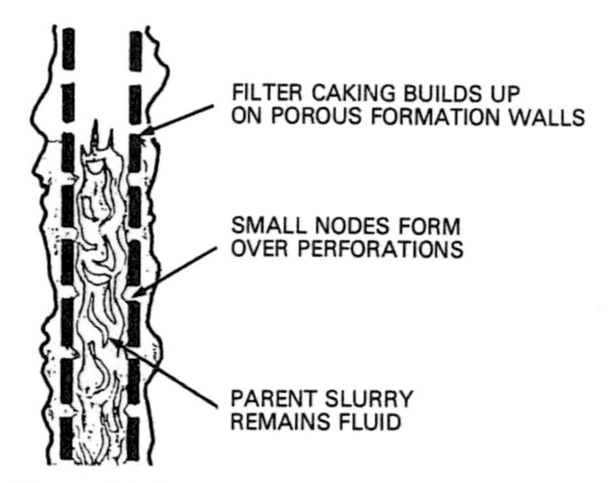

Figure 7.4. Low-pressure squeezing. From Dwight K. Smith, *Cementing* (Dallas: Society of Petroleum Engineers of AIME, 1976).

Placement Techniques

Two basic placement techniques can be used with a squeeze job, the bradenhead method and the packer method.

Bradenhead Method

The bradenhead method involves running tubing or drill pipe without a packer to a point below the perforations (fig. 7.5). With the hole full of water, cement is mixed and pumped down the tubing string to a predetermined height in the casing. The casing valve is open during this phase, and the water in the casing is circulated out the casing valve. When the cement is spotted over the perforation to the desired height in the

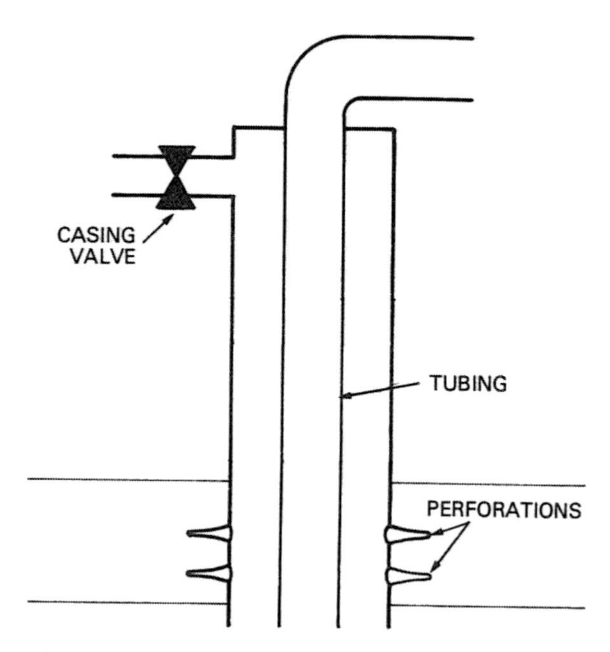

Figure 7.5. Bradenhead squeeze

casing, the casing valve is closed. The tubing is picked up out of the cement, and the desired squeeze pressure is applied by pumping down the casing. After the pressure is applied to the desired value, the excess cement can be reversed out by pumping water down the casing annulus and out the tubing string.

Since the full pressure is applied to the casing, the casing must be in good condition and of a weight and grade to withstand the pressures involved. This fact normally eliminates the bradenhead method for use in high-pressure squeeze jobs. A problem with the bradenhead method is that the hole must stand full of fluid in order to permit accurate spotting of the cement over the perforations.

Packer Method

Most squeeze jobs are performed using the packer method (fig. 7.1). In a high-pressure squeeze job, the packer is set as shown, and the formation is broken down by pumping water or mud down the tubing. Cement is then pumped down the tubing and out the perforations. When the desired surface pressure is obtained, the packer is unseated, and the excess cement is reversed out by pumping down the annulus.

A low-pressure squeeze is conducted in the same way except that the maximum surface pressure is held to a value that will result in a pressure at the perforations less than the fracture pressure of the formation.

In low-pressure squeeze jobs, difficulty is sometimes experienced in pumping the cement down if the packer is set before the cement is pumped. This difficulty is a special problem in tight formations, since the cement has to displace the water in the tubing string into the formation. It may not be possible in low-permeability formations to displace this water without fracturing the formations. If so, the packer is not set until the cement is on bottom. The volume of water in the tubing can be circulated up the casing annulus. When the cement is at the bottom of the tubing, pumping is stopped, the packer is set, and pumping is then continued.

Designing a Cement Squeeze Job

A number of factors must be considered in the design of a squeeze cement job. The correct slurry is essential.

Cement Slurry Design

Two important factors must be considered in the cement slurry design – enough pumping time, and the desired water-loss properties. The cement should remain in a fluid, pumpable state until the job is completed and any excess cement is reversed out. The filter loss should be low enough so that when pressure is applied and cement filtrate is lost to the formation, the cement will not flash-set.

Class A, G, or H cements can be used for squeeze cementing without additives to a depth of about 6,000 feet and will have adequate pumping times. Below 6,000 feet, a retarder is usually needed to obtain adequate pumping time. Table 7.1 shows some typical thickening times for Classes G and H cements with varying amounts of calcium lignosulfonate retarder. Such data can be used to pick the desired cement and retarder. Tests are normally run by using the actual well data before the squeeze job is performed.

TABLE 7.1
NORMALLY RECOMMENDED AMOUNTS AND THICKENING TIMES OF CALCIUM LIGNOSULFONATE RETARDER IN API CLASSES G AND H CEMENTS

	Temperature (°F)			
Depth (ft)	Static	Circulating	Retarder (percent)	Thickening Time (hours)
Casing Cementing				
4,000 to 6,000	140-170	103-113	0.0	3 to 4
6,000 to 10,000	170-230	113-144	0.0-0.3	3 to 4
10,000 to 14,000	230-290	144-206	0.3-0.6	2 to 4
14,000 to 18,000	290-350	206-300	0.6-1.0	*
Squeeze Cementing				
2,000 to 4,000	110-140	98-116	0.0	3 to 4
4,000 to 6,000	140-170	116-136	0.0-0.3	2 to 4
6,000 to 10,000	170-230	136-186	0.3-0.5	3 to 4
10,000 to 14,000	230-290	186-242	0.5-1.0	2 to 4*
14,000 plus	290 plus	242 plus	1.0 plus	*

*Requires special laboratory testing or the use of modified lignin retarder.

SOURCE: Dwight K. Smith, *Cementing* (Dallas: SPE-AIME, 1976)

Generally the fluid loss should be reduced to about 25–30 cc in 30 minutes on the standard API test. Tests should be run with the use of actual well data to be sure that the correct filtration-control additive is selected.

Fracture Gradient Considerations

When the high-pressure squeeze method was dominant, many rules of thumb circulated within the oil industry regarding the final surface pressure to use. With the high-pressure squeeze, enough pressure has to be exerted to break down the formation; this is done with a breakdown fluid before cement is pumped. The final squeeze pressure occurs when cement has set up in the casing; therefore, the final pressure on a high-pressure squeeze job may be irrelevant as long as the pressure does not rupture the casing.

The maximum surface pressure for a low-pressure squeeze, on the other hand, is very important, and the success of the job depends upon the maximum surface pressure being accurately determined. The job must also be conducted at a pressure less than the fracture pressure of the formation. Fracture pressures are evaluated by frac gradient data.

Frac gradient data. Frac gradients are usually expressed as psi/ft. A frac gradient of 0.6 psi/ft means that for every foot of depth, the pressure required to frac the formation is 0.6 psi. For example, a sand at 10,000 feet with a frac gradient of 0.6 psi/ft has a frac pressure of 6,000 psi. If the hole is filled with salt water with a pressure gradient of 0.5 psi/ft, a hydrostatic pressure of 5,000 psi will be exerted at the bottom of the hole. An applied surface pressure has

TABLE 7.2
BOTTOMHOLE TREATING PRESSURES AND FRACTURE GRADIENTS IN VARIOUS AREAS OF THE U. S.

Field	Formation	Depth (ft)	Bottomhole Treating Pressure (psi)	Fracture Gradient (psi/ft)
	Southwest Texas			
East Mathis	Frio	5,785-92	3,860	0.668
Cosden	Slick Wilcox	7,050	3,260	0.463
Seven Sisters	Argo	2,479-89	1,855	0.75
Stratton	Frio	6,226-30	4,225	0.68
Stratton	Frio	7,262-69	5,910	0.81
N. W. Freer	Wilcox Sand	6,814-25	4,980	0.73
Magnolia City	Frio Sand	5,626-33	3,630	0.645
	North Texas			
Grayback	Strawn	3,700	–	0.62
Electra	Cisco	1,600	–	1.07
Kamay	KMA	4,000	–	0.58
Graham	Strawn	3,121	–	0.69
North St. Jo	Strawn	2,400	–	0.58
Sherman	Davis	8,800	–	1.03
Corsicana	Wolf City	1,100	–	1.23
East Texas	Woodbine	3,600	–	0.725
Walnut Bend	Hudspeth	4,451	3,028	0.78
Dye Mound	Conglomerate	6,596	4,080	0.672
Burkburnett	Gunsight	1,550	1,824	1.14
Burkburnett	Canyon	2,140	1,874	0.88
Electra	Cisco	1,040	996	0.96
Kamay	KMA	3,900	2,660	0.68
	Rocky Mountain			
Elk Basin	Tensleep	6,100	4,395	0.72
Lost Cabin	Windriver	3,800	–	0.625
Garland	Basal Amsden	4,145	3,410	0.818
Red Wash (Uintah, Utah)	Green River	–	–	0.625
Red Desert (Sweetheart)	Almond (Mesa Verde)	–	–	0.80
Farmington-San Juan	Pictured Cliffs	2,340	1,565	0.67
Farmington-San Juan	Dakota	7,030	4,960	0.71
	West Texas			
Midland County	Lower Spraberry	8,180	4,330	0.529
Runnels County	Upper Gardner	3,870	1,570	0.41
	Oklahoma			
N. W. Stroud	Red Fork	3,485	2,200	0.631
Yale	Bartlesville	3,100	1,350	0.435
N. Bristow	Peru	2,350	1,100	0.468
Merrick	Wilcox	5,085	3,110	0.613
SE St. Louis	Earlsboro	3,206	2,165	0.675
Kiowa	Dolomite	775	890	1.15
E. Marlow	Helms	4,900	2,750	0.76
Sholem Alechem	Springer	5,400	3,800	0.70
Cement	Fortuna	2,352	1,930	0.82
Sholem Alechem	Springer	5,360	3,770	0.70
Lindsay	Hart	8,330	4,900	0.47
	Kansas			
Ellsworth County	Pennsylvanian Conglomerate	3,215	–	0.45
Rice County	Quartzite	3,125	–	1.31
McPherson County	Viola	3,383	–	0.64
Reno County	Simpson Sand	3,872	–	0.42

SOURCE: Shryock, S. H., and K. A. Slagle, "Problems Related to Squeeze Cementing," *J. Pet. Tech.* (Aug. 1968), 801–7.

to be less than 1,000 psi to prevent fracing the formation.

The frac gradient is independent of the depth of the formation. For example, the Slick-Wilcox formation in Southwest Texas has a gradient of 0.46 psi/ft at a depth of 7,000 feet. In North Texas, the Wolf City formation, found at a depth of 1,100 feet, has a frac gradient of 1.23 psi/ft. Many frac gradients are in the 0.6 psi/ft to 0.9 psi/ft range. Table 7.2 lists some typical fracture gradient data.

How frac gradient data are obtained. Frac gradient information is obtained from field data on frac jobs, squeeze jobs, injection well testing, or any other operation where fluid is pumped into a formation under pressure. A good example of how frac gradient data are accumulated is illustrated by step-rate tests run on injection wells in waterflood projects. It is usually undesirable to operate an injection well above the frac pressure of the formation because long fractures extending as much as several thousand feet can be created. Quite often these fracs intersect the wellbores of producing wells. The water is cycled through the fracture and does not move through the formation and displace the oil.

In order to prevent the problem, step-rate tests are run on new injection wells. Water is pumped into a well at constantly increasing rates, with the pressure being recorded at each rate. The data are plotted (fig. 7.6). As long as normal radial flow through the formation occurs, the rate is proportional to the difference between the wellbore pressure and the reservoir pressure. In the discussion of Darcy's radial flow equation, it was pointed out that flow is proportional to the pressure drawdown ($p_e - p_w$). The reverse is true in an injection well, where flow is proportional to ($p_w - p_e$). As long as the flow is radial, the data will plot as a straight line. As soon as a fracture is initiated, a rapid increase is realized in the rate, with very little pressure increase at the wellhead. The fracture pressure is the pressure at the break in the curve. In figure 7.6 the frac pressure is at a surface pressure of 1,800 psi. It is necessary to translate the surface pressure to bottomhole conditions to obtain the formation frac pressure.

Assume that the step-rate test of figure 7.6 was conducted in the well shown in figure 7.7. The pressure at the perforated interval is–

p_{BH} = surface pressure + hydrostatic pressure of 6,000' water column – friction loss (where depth of well is 6,000' and estimated friction loss is 60 psi)

$$p_{BH} = 1{,}800 \text{ psi} + 6{,}000' \times 0.052 \times 8.33 - 60 = 4{,}339 \text{ psi}$$

$$\text{Frac gradient} = \frac{4{,}339 \text{ psi}}{6{,}000 \text{ ft}} = 0.72 \text{ psi/ft.}$$

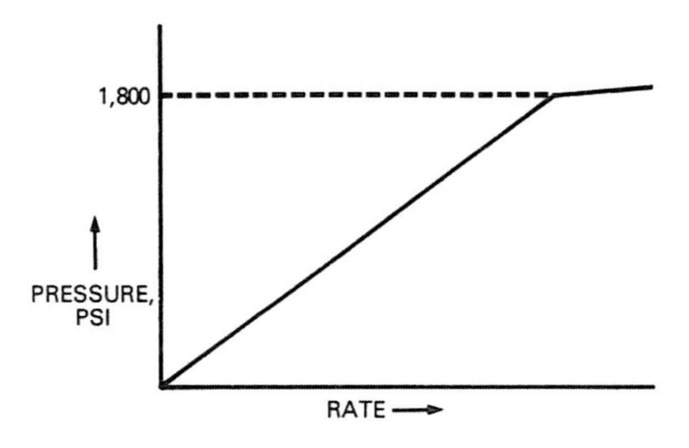

Figure 7.6. Step-rate test

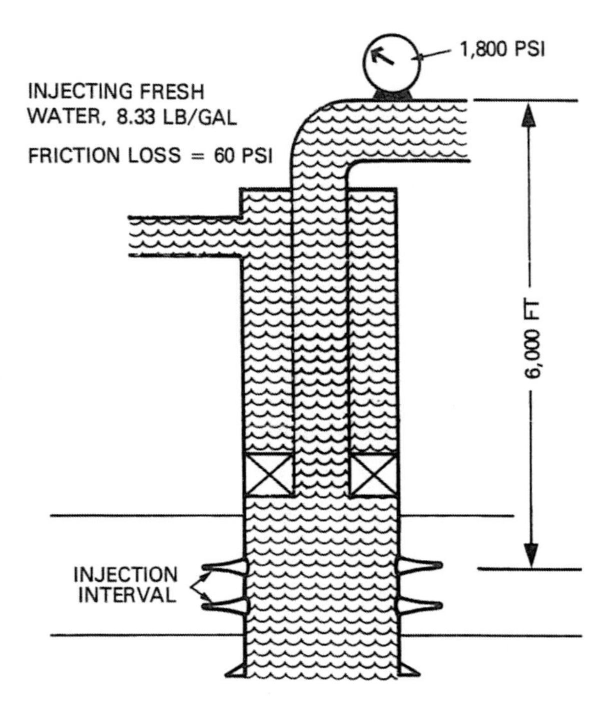

Figure 7.7. Step-rate test in a given well

The friction loss in a step-rate test is usually so low that it can be ignored. If the friction loss were eliminated from the above example, the frac gradient would be 0.73 psi/ft.

Data from frac jobs and high-pressure cement squeeze jobs are evaluated in a similar manner in order to obtain the frac gradient.

Maximum surface pressure calculations. Calculations to determine the maximum surface pressure on low-pressure squeezes are similar to those used in evaluating the step-rate test (fig. 7.6). The only complicating factor is that there are two different fluids, cement and water, making up the hydrostatic head. The following example problem illustrates the principles involved.

It is desired to squeeze perforations at 6,970–80' through 2⅞" 6.5-lb/ft tubing with a squeeze packer set at 6,950', using 75 sacks of Class H cement. The cement will be displaced with 9.0-lb/gal salt water. The annulus is loaded with 9.0-lb/gal salt water. The maximum surface pressure should have a safety factor of 300 psi to avoid fracturing the formation (fig. 7.8).

NOTE: Assume that the cement fills the casing to the base of the perforations.

Frac gradient = 0.75 psi/ft
Cement yield = 1.05 cu ft/sack
Cement weight = 16.5 lb/gal

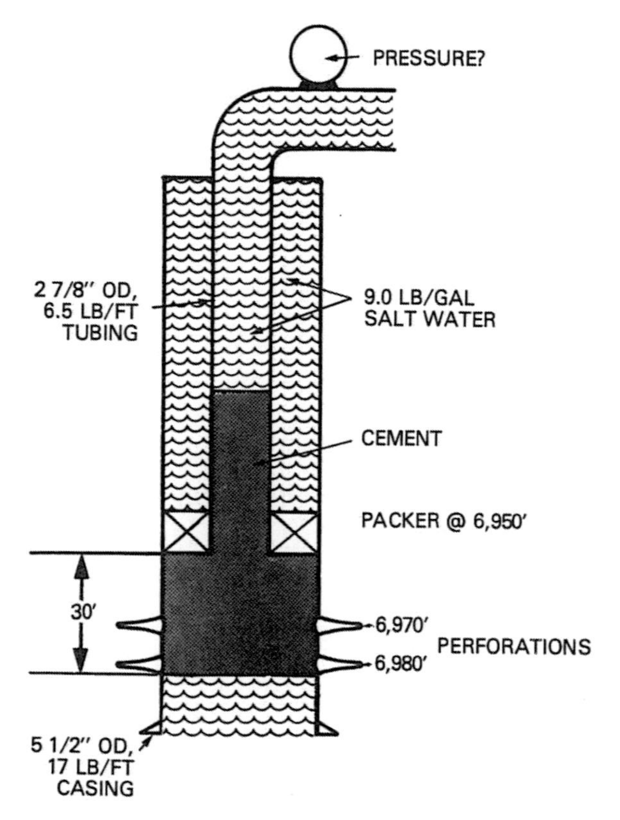

Figure 7.8. Calculating maximum surface pressure for squeeze job

The problem is to find out how much surface pressure can be applied without fracing the formation.

First, calculate the fracture pressure at the midpoint of the perforations:

$$p_{frac} = 6{,}975 \text{ ft} \times 0.75 \text{ psi/ft} = 5{,}231 \text{ psi.}$$

The combination of surface pressure, hydrostatic pressure,and safety factor cannot exceed 5,231 psi.

Next, calculate the slurry volume.

Slurry volume = 75 sacks × 1.05 cu ft/sack
Slurry volume = 78.8 cu ft.

Part of the cement is in the casing, and part is in the tubing. First, calculate the amount in the casing. Note that there are 30 feet of cement in the casing. From a service company handbook, the capacity of 5 ½" OD, 17-lb/ft casing is found to be 0.1305 cu ft/linear foot.

Volume of cement in casing = 30' × 0.1305 cu ft/lin ft = 3.9 cu ft

Volume of cement in tubing = slurry volume − volume in casing = 78.8 cu ft − 3.9 cu ft = 74.9 cu ft.

From a service company handbook, the capacity of 2⅞" OD, 6.5-lb/ft tubing is found to be 0.0325 cu ft/lin ft.

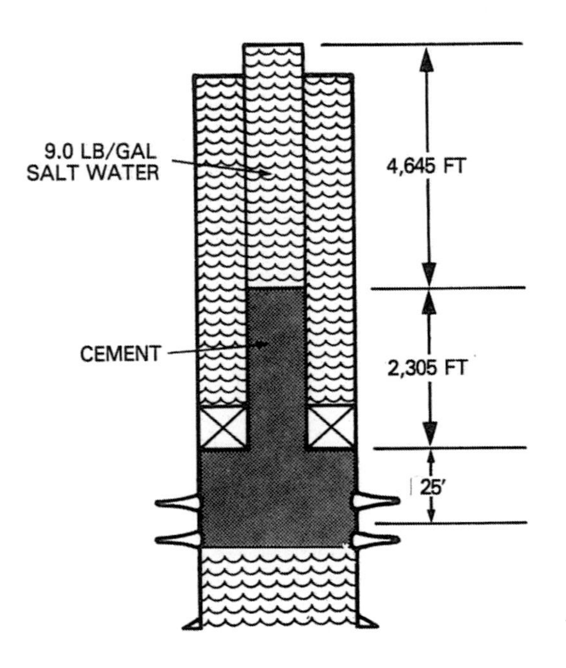

NOTE: The hydrostatic head is calculated at the midpoint of the perforations, since that is where the frac pressure was calculated.

Figure 7.9. Calculating maximum surface pressure for squeeze job

$$\text{Height of cement in tubing} = \frac{74.9 \text{ cu ft}}{0.0325 \text{ cu ft/lin ft}} = 2{,}305 \text{ ft.}$$

The data that have been calculated are shown in figure 7.9.

NOTE: The hydrostatic head was figured at the midpoint of the perforations, since that is where the frac pressure was calculated.

Hydrostatic head (HH) = 2,330' of 16.5 lb/gal cement + 4,645' of 9.0-lb/gal water.

HH = 2,330' × 16.5 × 0.052 + 4.645 × 9.0 × 0.052

HH = 4,173 psi.

Safety factor = 300 psi

Total = 4,473 psi.

Frac pressure = 5,231 psi

Maximum surface pressure = 5,231 − 4,473 = 758 psi.

NOTE: Friction loss is ignored, since fracture of the formation would occur at zero flow rate.

The calculations indicate that a low-pressure squeeze job can be performed satisfactorily on the example well if the maximum surface pressure is limited to 758 psi.

Block Squeeze

A common practice in some areas is to perform a high-pressure block squeeze by perforating the zone below the producing interval and then performing a high-pressure squeeze. The zone above the producing interval is then perforated and squeezed off in a similar manner. The hole is drilled out, and the producing interval is perforated. The purpose of block squeezing is to isolate the producing interval and prevent communication with the sands immediately above and below the producing interval (fig. 7.10). The technique is very questionable. It has been pointed out that the formation is fractured vertically when cement is pumped into it under high pressure. The vertical fracture may allow the intervals above and below the zone being squeezed off to communicate. Since the goal of a block squeeze is to prevent communication, the process is self-defeating.

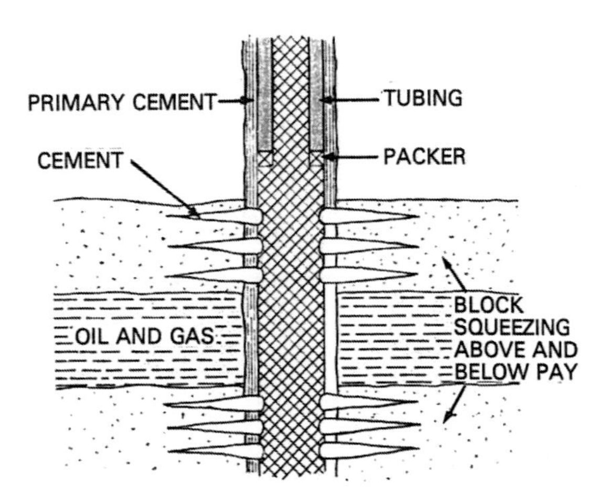

Figure 7.10. Block squeeze. From Dwight K. Smith, *Cementing* (Dallas: Society of Petroleum Engineers of AIME, 1976)

Another bad aspect of block squeezes is that there are squeezed-off perforations above the producing interval. Such squeezed-off perforations can break down and cause difficult well problems. Experience has shown that the best rule to follow in depleting a multipay well is to start at the bottom and, moving upward, deplete each zone. Squeezed-off perforations that may break down are thus avoided. Some very expensive workovers have been caused by the breakdown of perforations above the producing zone.

Circulation Squeeze

The circulation squeeze is an interesting variation of the squeeze cementing process. It is applicable where there is communication behind the pipe because of channeling of the cement or where there is essentially no cement in the annulus (fig. 7.11). In the figure, gas from an upper zone has channeled through a poor cement job into a lower oil-producing zone.

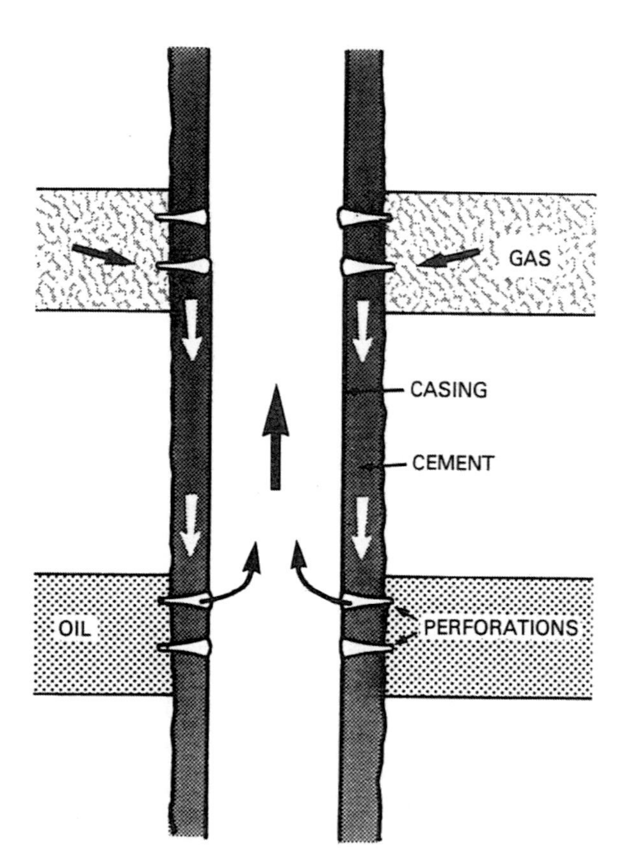

Figure 7.11. Communication between zones behind pipe

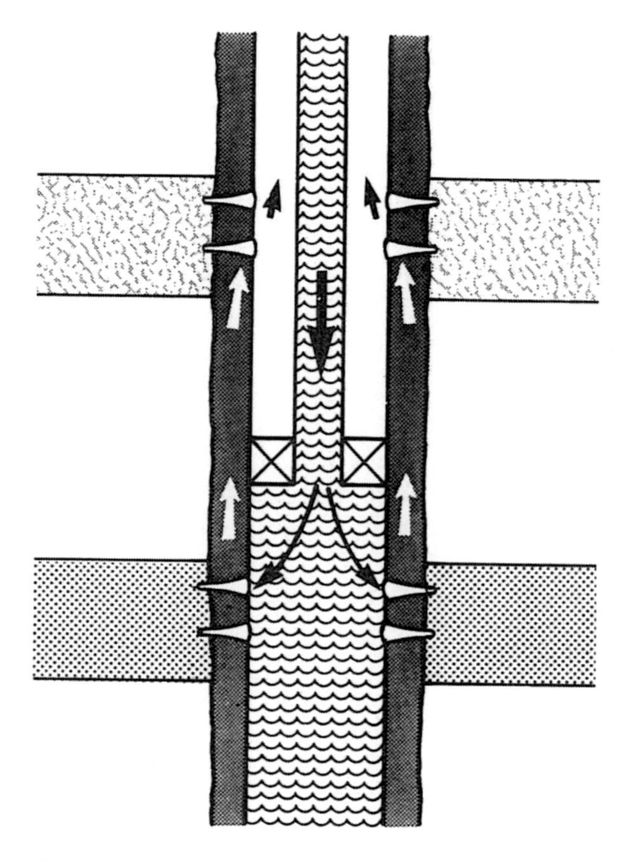

Figure 7.12. Circulation cement squeeze

The first step in the circulation squeeze (fig. 7.12) is to perforate the upper gas sand. Next, tubing is run with a packer, and the packer is set between the two perforated intervals. Water is then circulated between the two zones to remove as much mud as possible from the channel. Cement is then pumped through the channel. After the cement is circulated, the packer is released and picked up above the upper perforation, a low squeeze pressure is applied, and the excess cement is circulated out.

There is no good method to estimate the amount of cement to leave behind the pipe. This amount is not critical except that if an excess amount is pumped, it will collect on top of the packer. If the cement slurry has sufficient pumping time, no problem should exist.

VIII Packer and Tubing Forces

Packers are devices that are run on tubing to make a seal between the casing and the tubing. They are run either at the time of the original completion or during a workover. After the packer is run and the well is placed on production, the pressure and temperature in the tubing string change and cause tubing movement or create forces in the tubing-packer system. Also when the well is acidized or fraced, pressure and temperature in the tubing string change and create forces in the packer-tubing system. The forces induced need to be evaluated *before* the packer is run, since they may cause a retrievable packer to unseat or the tubing seals to pull out of a permanent packer.

Types of Packers

Packers are of two basic types–retrievable and permanent. The packer body, slips, and rubber packing elements are rigidly attached to the tubing string in the retrievable type. In the permanent type, the packer body with slips and rubber packing elements are usually run and set on wireline, and then the tubing with seal subs is stabbed into the packer to make a pressure-tight connection.

Retrievable Packers

Retrievable packers can be divided into three general categories: (1) weight-set, (2) tension-set, and (3) mechanically or hydraulically set.

Weight-set packers. The simplest type of retrievable packer is the weight-set type. The packer has a set of slips with teeth that point in a downward direction. It is run to the desired depth, with the slips retracted (fig. 8.1). At the desired setting depth, the slips are engaged, usually by rotating the tubing and disengaging a pin from a J-slot. The packer is set by applying tubing weight. When this is done, the cone presses outward on the slips and causes them to bite into the casing wall. The packer rubber sealing element is compressed by the tubing weight that is applied, and the annulus is sealed off. The slips resist movement in the downward direction only.

Weight-set packers (fig. 8.2) are the cheapest type and are usually run in low-pressure applications such as production packers.

Tension-set packers. Tension-set packers are basically weight-set packers run upside down. They operate in the same manner as weight-set packers except that they are set by picking up on the tubing and placing it in tension. Figure 8.3 is a schematic drawing of a tension-set packer. A typical application for a tension-set packer is one in shallow injection wells where the pressure in the tubing at the packer might cause a weight-set packer to unseat.

Mechanically set or hydraulically set packers. Weight-set and tension-set packers were the first types to be developed. Subsequently, packers that are set by rotation of the tubing or by the application of hydraulic pressure inside the tubing have been developed.

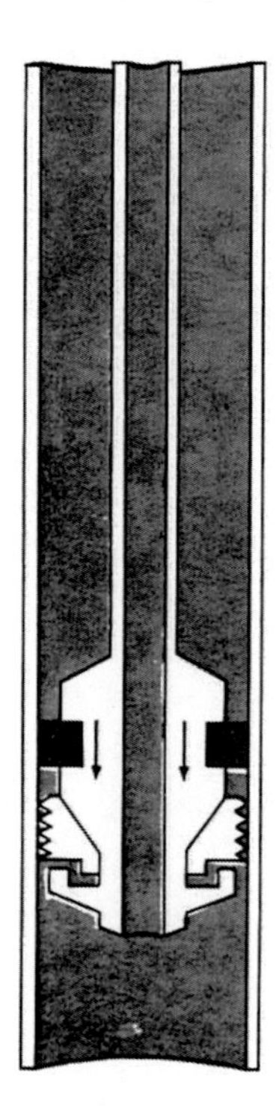

Figure 8.1. Weight-set packer with slips retracted (Courtesy of Baker Packers)

Figure 8.2. Weight-set packer (Courtesy of Baker Packers)

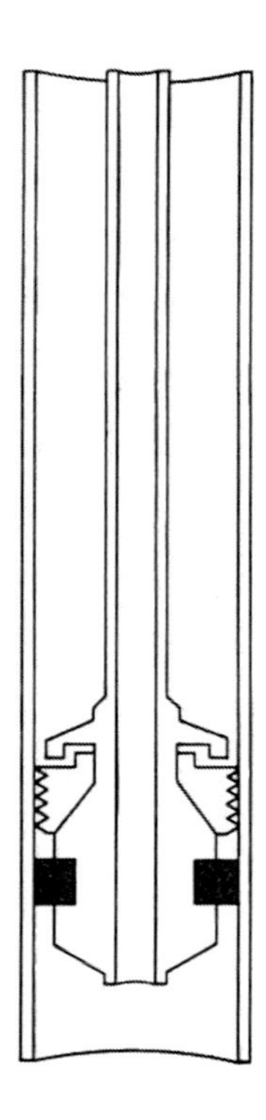

Figure 8.3. Tension-set packer after being set (Courtesy of Baker Packers)

Figure 8.4. Mechanically set packer (Courtesy of Baker Packers)

These packers utilize neutral slips that resist movement in either direction. A typical mechanically set packer is shown in figure 8.4. Information can be obtained from packer company catalogs regarding these packers.

Permanent Packers

The body of a permanent packer, including the slips and the sealing element, are run and set on a wireline or on tubing. The tubing with seal subs and locator or latch unit is run and the seal subs seated in the smooth bore of the packer (fig. 8.5).

The packer body has two sets of slips and a rubber sealing unit (fig. 8.5*A*). The packer body is run to the desired depth on a setting tool. A charge is fired, and the setting tool pushes down on the top of the packer and up on the bottom. This action compresses the packing element and sets the packer body slips. The upper slips resist upward movement, and the lower slips resist downward movement. The packing element is retained in an expanded condition by the two sets of slips.

The setting tool that is used when the packer body is set on wireline utilizes an explosive charge to supply the setting force. When the packer body is set on tubing, hydraulic pressure inside the tubing supplies the setting force.

The tubing is run with a locator or latch sub and seal subs (fig. 8.5*A*). The seal subs are accurately machined to fit into the smooth internal bore of the packer. They have rubber sealing elements to seal against the packer bore. The locator sub is larger than the packer bore and is used to locate the top of the packer. The latch sub helps locate the top of the packer but also has a set of segmented threads that seat in the top of the packer and prevent upward movement of the sealing elements. Figure 8.5*B* shows the packer with the seal subs in place and the well ready for production or treatment.

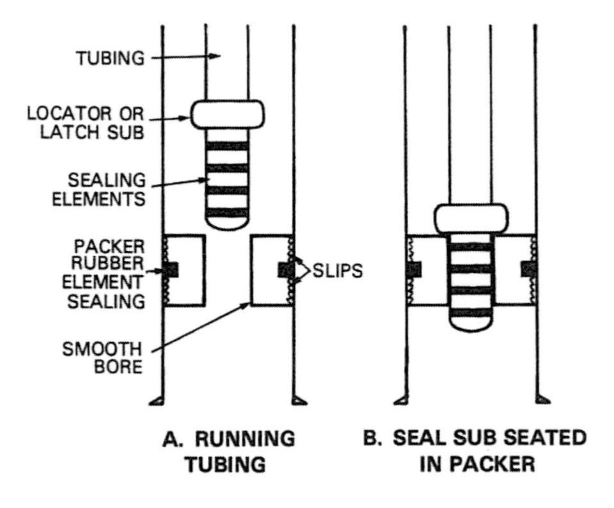

Figure 8.5. Permanent packer

Permanent packers are so named because they cannot be retrieved and must be drilled up if a change in setting is desired.

Forces Acting on Packer-Tubing Systems

Tubing movement occurs, or forces are induced into the packer-tubing system, when a well is placed on production or treated by fluids being pumped down the tubing string. Each change in status must be evaluated separately to ensure that the operation can be carried on safely, and each change in status is evaluated against the static conditions before the packer was run. For example, the following changes in status may be made for a well:

1. Running a packer
2. Placing the well on production
3. Swabbing the well down
4. Acidizing the well
5. Fracing the well

After a packer is run, each of the subsequent operations is evaluated against the original conditions that prevailed at the time the packer was set. For example, a frac job is the last operation listed. The frac job conditions must be evaluated against the original conditions that existed at the time the packer was run. The forces that must be considered follow.

Forces Acting on Packer Body

Pressure changes in the tubing or casing annulus will result in forces being applied to the packer body. The pressure is the same in the tubing and the annulus when the packer is set,

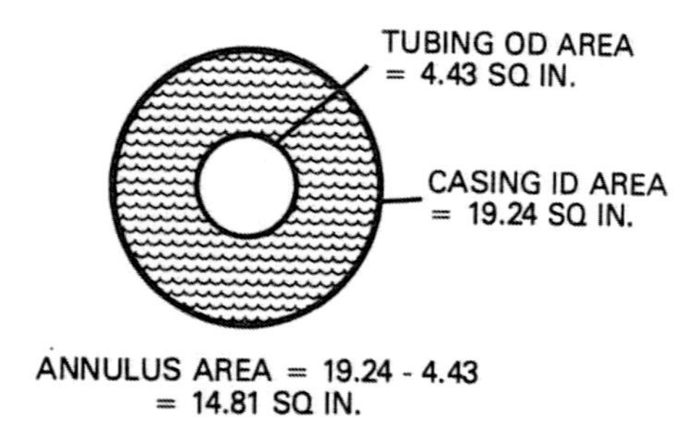

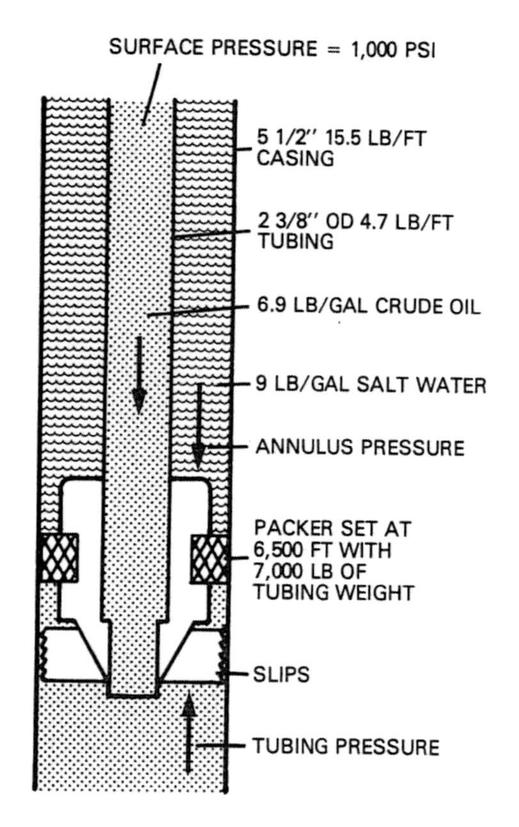

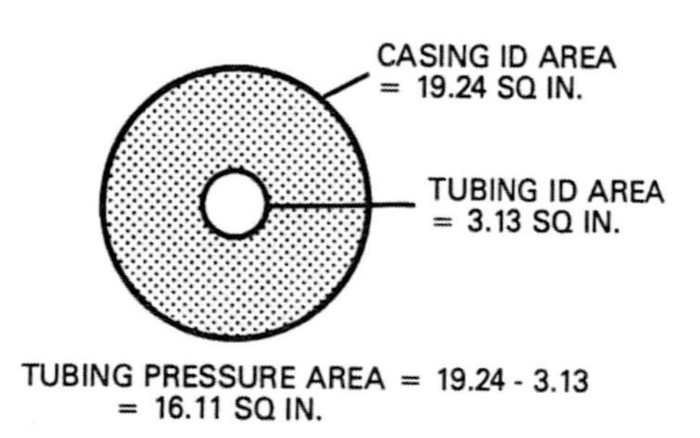

Figure 8.6. Weight-set packer force balance for acidizing

since the hole is full of kill fluid. Pressure inside the tubing changes if the well is placed on production or if the well is treated. The forces due to pressure changes in the tubing are more critical for weight-set packers, and they can be considered first.

Retrievable weight-set packers. Retrievable weight-set packers are commonly used as production packers. In production service the weight-set packer is generally satisfactory, since the pressure below the packer is decreased in order for the well to produce. If it is desired to perform any type of treatment such as acidizing on a well with a weight-set packer when the pressure below the packer is increased, calculations must be made to determine whether the packer will become unseated. The calculations can best be illustrated by the following example problem.

Assume that a well with a weight-set packer is to be acidized (fig. 8.6). It is common practice to make a force-balance calculation on the packer body to see if it will unseat. Note that the packer is held down by the amount of tubing weight, 7,000 lb, that is placed on it and by the annular pressure acting on its top. The tubing pressure acts on the bottom cross-sectional area of the packer, which is 16.11 square inches; this is larger than the cross-sectional annular area above the packer, which is only 14.81 square inches. When the packer is initially set, the pressure of the kill fluid at the packer is, for practical purposes, the same above and below the packer. Since there is a larger area exposed to the kill fluid pressure below the packer, there is an upward force. This is the buoyancy effect.

A simple force-balance calculation for the packer during acidizing operations is made as follows:

Hydrostatic pressure at packer:

Annulus = 9.0 lb/gal × 0.052 × 6,500 ft
= 3,042 psi

Tubing = 6.9 lb/gal × 0.052 × 6,500 ft
= 2,332 psi

Total pressure at packer:

Annulus = 3,042 psi

Tubing = 1,000 psi (surface) + 2,332 psi
= 3,332 psi

Downward forces:

Tubing weight on packer		= 7,000 lb
Annular force = 3,042 psi × 14.81 sq in.		= 45,050 lb
	Total	= 52,050 lb

Upward forces:

Tubing pressure = 3,332 psi × 16.11 sq in. = 53,680 lb

The resultant upward force is 1,630 lb (53,680 − 52,050). Obviously, if the well is acidized under the conditions shown in the figure, the packer will unseat. The well may still be acidized if corrective steps are taken to increase the downward forces.

One way to increase the downward force is to add more tubing weight, but this would not be desirable, since the entire tubing string weight is only 30,550 lb. In addition, the tubing would tend to buckle mechanically from the added weight, and very little of the desired weight would reach the packer.

The best way to increase the downward force is to put pressure on the casing annulus. By applying a pressure of 583 psi to the annulus, the original 7,000 lb of resultant downward force on the packer can be restored.

Additional force = 583 psi × 14.81 sq in. = 8,630 lb

Resultant downward force on packer = 8,630 lb − 1,630 lb = 7,000 lb

The well can then be acidized if a pressure of 583 psi is applied to the annulus. Adding such a pressure should present no problem, provided that the casing is known to be in good condition and can stand the surface pressure of 583 psi.

The simple force-balance evaluation is usually sufficient, but it may result in problems if extremely cold fluids are pumped down the tubing.

Permanent packers. The forces acting on the body of a permanently set packer are normally ignored. In most instances, a permanent packer will resist movement from any pressure differential that can be applied without rupturing the casing. For extreme conditions, the manufacturer's catalog should be consulted.

Forces Acting on Tubing

Pressure-temperature changes can cause forces or length changes in the tubing string. Four effects must be evaluated:

1. Piston effect
2. Pressure, or helical, buckling effect
3. Ballooning effect
4. Temperature effect

Since the piston effect and tubing movement due to the pressure buckling effect are mainly concerned with permanent packers, the discussion of these effects will be limited to permanent packer installations. Later, the effects that should be considered in retrievable packer installations will be discussed.

Hooke's law. If a piece of steel, such as tubing, is stressed by applying a force to it, it will change in length. If the force is released, the tubing will return to its original length. Every material, such as steel, has an elastic limit, and if the elastic limit is exceeded, the material will not return to its original length. Hooke's law is valid below the elastic limit.

Robert Hooke, an English physicist and mathematician in the 17th century, found that strain is proportional to stress.

Stress = force/cross-sectional area = F/A

Strain = change in length/total length = $\Delta L/L$

Hooke found that strain is proportional to stress. In other words, a plot of stress vs. strain results in a straight line (fig. 8.7). Thomas

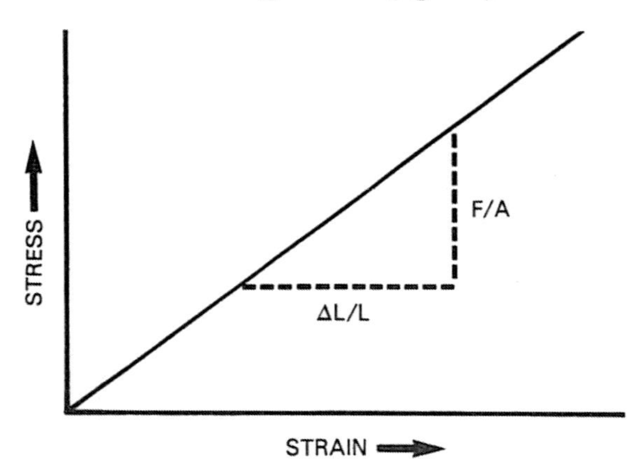

Figure 8.7. Stress versus strain

Young found that the slope of the line is constant for a given material, or stress/strain is equal to a term he called the *modulus of elasticity*. Young's modulus of elasticity happens to be 30,000,000 psi for steel. It is denoted by the letter E.

$$\text{Stress/strain} = E$$
$$\text{Stress} = F/A$$
$$\text{Strain} = \Delta L/L$$
$$\frac{F/A}{\Delta L/L} = E$$

Simplifying,

$$F = \frac{\Delta LEA}{L} \quad \text{or}$$

$$\Delta L = \frac{FL}{EA}$$

The force equation can be used to determine how many pounds to pull on a tubing string to get a desired amount of stretch. The ΔL equation can be used to determine how much the tubing will stretch if a force is applied. Since cross-sectional areas of tubing are normally given in square inches and tubing lengths in feet, it is necessary to convert the length in feet to inches when Hooke's law is used in the above form.

Piston effect. When the seal subs on a tubing string are seated in a permanent packer, a piston is created (fig. 8.8). Without a locator or latch sub, the seal subs can move up or down in the packer bore like the piston in an automobile engine.

The following nomenclature applies:

A_o = area of tubing OD (outside diameter)
A_i = area of tubing ID (inside diameter)
A_p = area of packer bore
p_o = pressure in annulus at packer
p_i = pressure in tubing at packer

Note that the annulus pressure, p_o, acts on the area $A_p - A_o$, and the tubing pressure, p_i, acts

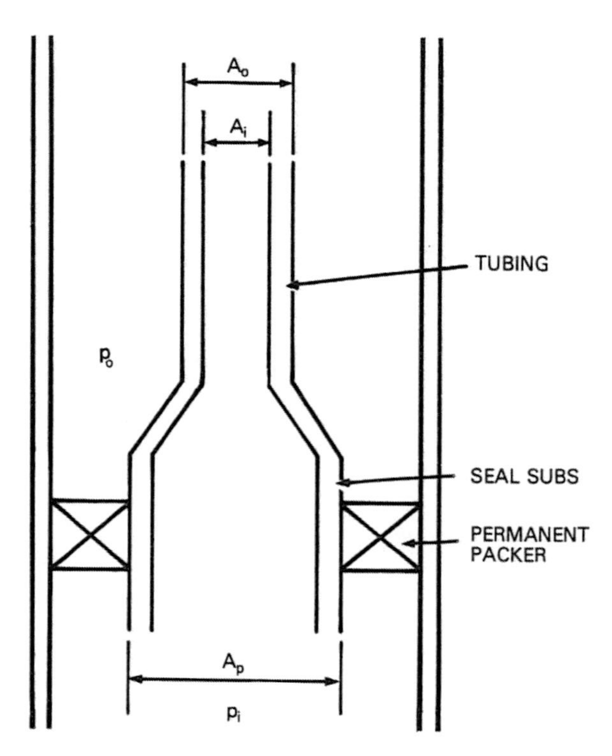

Figure 8.8. Piston effect

on the area $A_p - A_i$. When the tubing is set in the packer, the hole is normally loaded with fluid, so the same pressure exists in the tubing and the annulus at the packer. This condition exists when the tubing is landed. Packer-tubing calculations are concerned with the changes that occur after the tubing is landed. For example, assume that there is 9.0-lb/gal salt water in the hole when the tubing is run. When production begins, the salt water is swabbed from the tubing, and the pressure in the tubing drops, so for most producing conditions the pressure in the tubing at the packer, p_i, is less than the pressure in the annulus, p_o, at the packer. If the well is stimulated by pumping fluids down the tubing, p_i will normally be greater than p_o.

Since the changes from the original condition are of concern, Δp_i and Δp_o will be used to designate the changes in pressure. Since a pressure acting on a cross-sectional area results in a force, the following formula for piston effect can be obtained.

$$F_1 = (A_p - A_i)\Delta p_i - (A_p - A_o)\Delta P_o$$

where

F_1 = force exerted on the tubing due to the piston effect.

The force on the tubing will cause the tubing either to contract or to elongate, depending on the magnitude of Δp_i and Δp_o. Tubing will stretch or contract in accordance with Hooke's law.

$$\Delta L = \frac{LF}{EA_s}$$

where

E = modulus of elasticity
= 30,000,000 psi for steel
A_s = cross-sectional area of tubing.

Substituting F_1 for F in Hooke's law, the following equation for the change in tubing length is obtained:

$$\Delta L_1 = \frac{-L}{EA_s}\left[(A_p - A_i)\Delta p_i - (A_p - A_o)\Delta p_o\right]$$

The use of the piston effect formula can best be illustrated by the following example problem.

An acid job is to be performed on the well shown in figure 8.9.

Tubing OD = 2.875 in. A_o = 6.492 sq in.
Tubing ID = 2.441 in. A_i = 4.68 sq in.
Tubing cross-sectional area, A_s = 6.492 − 4.68
= 1.812 sq in.
Packer seal bore = 4.0 in. A_p = 12.567 sq in.

The well will be acidized using an estimated surface tubing pressure of 6,000 psi, and 2,500-psi surface pressure will be placed on the casing annulus.

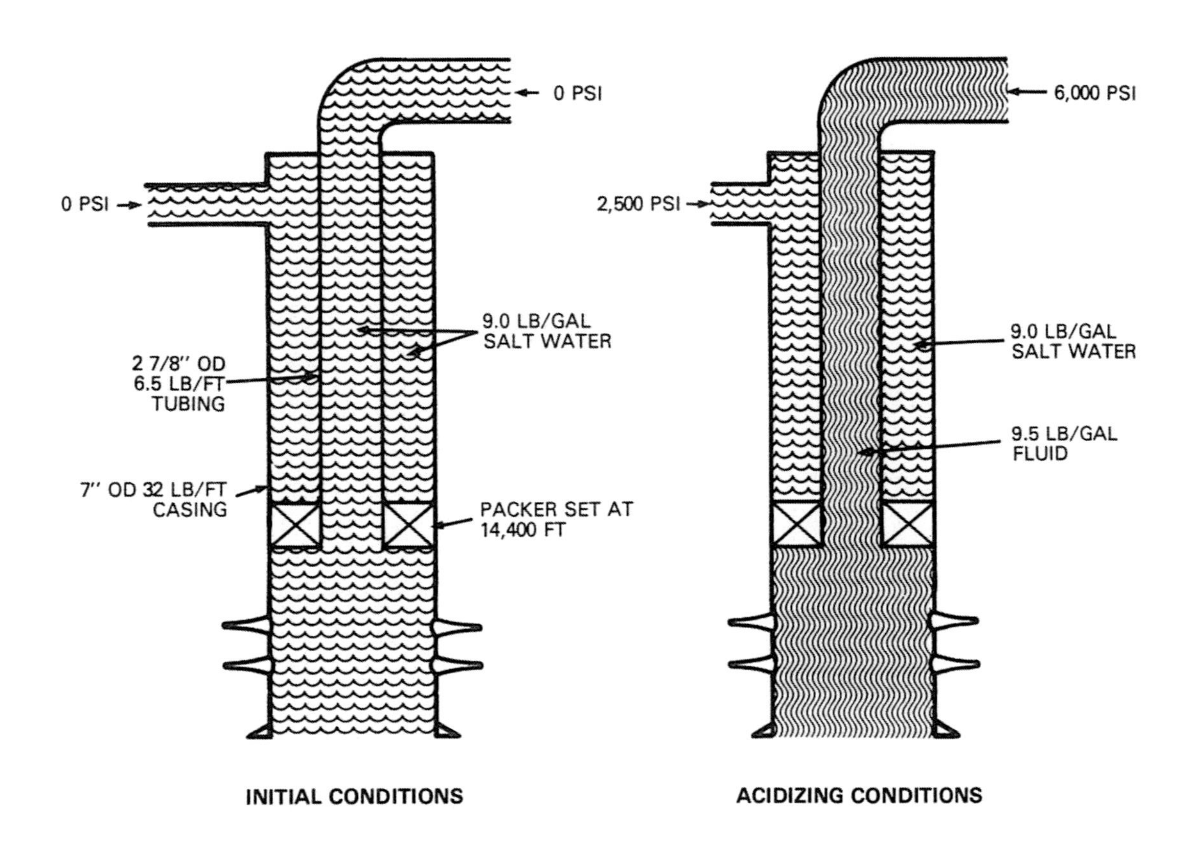

Figure 8.9. Example problem—permanent packer

The first step is to calculate the values of Δp_i and Δp_o as follows. Remember that p_i and p_o are the pressures in the tubing and the annulus at the packer.

Initial Conditions

$$p_i = p_o = 14{,}400 \text{ ft} \times 9.0 \text{ lb/gal} \times 0.052 = 6{,}739 \text{ psi}$$

(lb/gal × 0.052 = psi/ft)

Acidizing Conditions

$$p_i = 6{,}000 \text{ psi} + 14{,}400 \text{ ft} \times 9.5 \text{ lb/gal} \times 0.052 = 13{,}114 \text{ psi}$$

$$p_o = 2{,}500 \text{ psi} + 14{,}400 \text{ ft} \times 9.0 \text{ lb/gal} \times 0.052 = 9{,}239 \text{ psi}$$

$$\Delta p_i = \text{final } p_i - \text{initial } p_i = 13{,}114 - 6{,}739 = 6{,}375 \text{ psi}$$

$$\Delta p_o = \text{final } p_o - \text{initial } p_o = 9{,}239 - 6{,}739 = 2{,}500 \text{ psi}$$

$$\Delta L_1 = \frac{-L}{EA_s}[(A_p - A_i)\Delta p_i - (A_p - A_o)\Delta p_o]$$

$$\Delta L_1 = -\frac{12 \times 14{,}400}{30 \times 10^6 \times 1.812}[(12.567 - 4.68) \times 6{,}375 - (12.567 - 6.492) \times 2.500]$$

$$\Delta L_1 = -111.6 \text{ inches.}$$

Note that length is converted to inches to be consistent, since tubing and packer bore measurements are normally given in inches.

The tubing will contract 111.6 inches, or about 9 feet, due to the piston effect. The minus sign indicates an upward movement and the plus sign a downward movement if the sign convention outlined in figure 8.10 is used.

Helical buckling. Helical buckling occurs when the pressure inside the tubing at the packer is greater than the pressure in the annulus at the packer. The higher pressure inside the tubing causes the tubing to buckle, or "corkscrew". Helical, or pressure, buckling should not be confused with mechanical buckling caused by slacking off too much weight on the tubing.

Lubinski et al.[1] developed the following formula to calculate length changes in the tubing due to helical buckling:

$$\Delta L_2 = \frac{-r^2 A_p^2(\Delta p_i - \Delta p_o)^2}{8EI(W_s + W_i - W_o)}$$

where

r = radial clearance between casing and tubing, inches

I = moment of inertia for tubing $= \pi/64(OD^4 - ID^4)$, inches^4

W_s = weight of tubing, lb/in.

W_i = weight of fluid inside tubing, lb/in.

W_o = weight of fluid outside tubing, lb/in.

All other terms are those used in the piston-effect equation.

The effects of helical buckling can be demonstrated by continuing the example problem given in figure 8.9.

Length Change

Negative length changes refer to upward tubing movement.

Positive length changes refer to downward tubing movement.

Force

Negative forces refer to tension.

Positive forces refer to compression.

Pressure Changes

Negative pressure changes refer to pressure reduction.

Positive pressure changes refer to pressure increase.

$$p = p_{final} - p_{initial}$$

Temperature Changes

Negative temperature changes refer to temperature reduction.

Positive temperature changes refer to temperature increase.

$$t = t_{final} - t_{initial}$$

Figure 8.10. Sign convention (Courtesy of Baker Packers)

1. Arthur Lubinski, W. H. Althouse, Jr., and J. L. Logan, "Helical Buckling of Tubing Sealed in Packers," *Journal of Petroleum Technology*, June, 1962.

TABLE 8.1
$W_s + W_i - W_o$

Tubing OD (inches)	Weight (lb/in.)	W_i and W_o (lb/in.)	7.0	8.0	9.0	10.0	11.0	12.0	13.0	14.0	15.0	16.0	17.0	18.0	Lb/Gal
			52.3	59.8	67.3	74.8	82.3	89.8	97.2	104.7	112.2	119.7	127.2	134.6	Lb/Cu Ft
1.66	W_s = .200	W_i	.045	.052	.058	.065	.071	.078	.084	.091	.097	.104	.110	.116	
		W_o	.065	.075	.084	.094	.103	.112	.122	.131	.140	.150	.159	.169	
1.90	W_s = .242	W_i	.062	.070	.079	.088	.097	.106	.115	.123	.132	.141	.150	.159	
		W_o	.086	.098	.110	.123	.135	.147	.159	.172	.184	.196	.209	.221	
2.00	W_s = .283	W_i	.066	.076	.085	.095	.104	.114	.123	.133	.142	.152	.161	.171	
		W_o	.095	.109	.122	.136	.150	.163	.177	.190	.204	.218	.231	.245	
2 1/16	W_s = .283	W_i	.073	.083	.094	.104	.114	.125	.135	.146	.156	.167	.177	.187	
		W_o	.101	.116	.130	.145	.159	.174	.188	.202	.217	.231	.246	.260	
2 3/8	W_s = .392	W_i	.095	.108	.122	.135	.149	.162	.176	.189	.203	.217	.230	.243	
		W_o	.134	.153	.172	.192	.211	.230	.249	.268	.288	.307	.326	.345	
2 7/8	W_s = .542	W_i	.142	.162	.182	.203	.223	.243	.263	.284	.304	.324	.344	.364	
		W_o	.196	.225	.253	.281	.309	.337	.365	.393	.421	.450	.478	.506	
3 1/2	W_s = .767	W_i	.213	.243	.274	.304	.335	.365	.395	.426	.456	.487	.517	.548	
		W_o	.291	.333	.365	.416	.458	.500	.541	.583	.625	.666	.708	.749	

Weight of steel $W_s = \frac{\text{Pipe weight (lb/ft)}}{12}$

Weight of fluid in tubing $W_i = \frac{\text{Mud weight (lb/gal)}}{231} \times A_i$

Weight of displaced fluid $W_o = \frac{\text{Mud weight (lb/gal)}}{231} \times A_o$

SOURCE: Baker Packers, *Packer Calculations Handbook,* p. 108.

Given:

Casing ID = 6.094 inches

$$r = \frac{6.094 - 2.875}{2} = 1.61 \text{ inches}$$

$$I = \frac{\pi}{64}(2.875^4 - 2.441^4) = 1.611 \text{ inches}^4$$

$$\Delta L_2 = -\frac{r^2 A_p^2}{8EI}\frac{(\Delta p_i - \Delta p_o)^2}{(W_s + W_i - W_o)}$$

$$\Delta L_2 = \frac{-1.61^2 \cdot 12.567^2(6{,}375 - 2{,}500)^2}{8 \cdot 30 \cdot 10^6 \cdot 1.611(0.542 + 0.192 - 0.253)}$$

$$= -33.1 \text{ inches.}$$

Values of W_s, W_i, and W_o are from table 8.1.

The tubing will contract 33.1 inches due to helical buckling. This is in addition to the 111.6 inches of contraction due to the piston effect. The upward movement is indicated by the minus sign, as previously explained.

Ballooning effect. The third effect is ballooning. When the pressure inside the tubing is greater than it is on the outside, the tubing will tend to blow up like a balloon and shorten (fig. 8.11*A*). If the pressure is greater on the outside, the result is reverse ballooning (fig. 8.11*B*). Reverse ballooning causes the tubing to elongate.

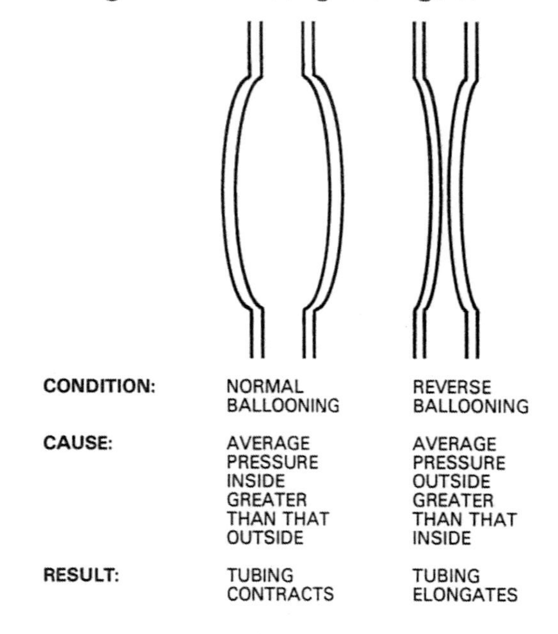

Figure 8.11. Balloon effect

Ballooning is described by the following equation:

$$\Delta L_3 = \frac{-2L\gamma}{E}\left[\frac{\Delta p_{ia} - R^2\Delta p_{oa}}{R^2 - 1}\right]$$

The following new terms have been introduced:

γ = Poisson's ratio = 0.3 for steel
Δp_{ia} = change in *average* pressure in the tubing
Δp_{oa} = change in *average* pressure in the annulus
R = tubing OD/tubing ID.

The effects of ballooning are demonstrated by continuing the example problem of figure 8.9.

$$R = \frac{\text{tubing OD}}{\text{tubing ID}} = \frac{2.875}{2.441} = 1.178$$

$R^2 = 1.387.$

The average pressure is simply the surface pressure plus the bottomhole pressure divided by 2.

For initial conditions:

$$p_{ia} = p_{oa} = \frac{0 + 6{,}739}{2} = 3{,}370 \text{ psi}$$

For treating conditions:

$$p_{ia} = \frac{6{,}000 + 13{,}114}{2} = 9{,}557 \text{ psi}$$

$$p_{oa} = \frac{2{,}500 + 9{,}239}{2} = 5{,}870 \text{ psi}$$

$$\Delta p_{ia} = 9{,}557 - 3{,}370 = 6{,}187 \text{ psi}$$

$$\Delta p_{oa} = 5{,}870 - 3{,}370 = 2{,}500 \text{ psi}$$

$$\Delta L_3 = \frac{-2 \times 14{,}400 \times 12 \times 0.3}{30{,}000{,}000} \times \frac{6{,}187 - 1.387 \times 2{,}500}{1.387 - 1}$$

$\Delta L_3 = -24.3$ inches.

The tubing will move upward 24.3 inches as a result of the ballooning effect.

Temperature effect. The temperature is the simplest of the four tubing effects to evaluate. All that is necessary is to determine the average temperature change and multiply it by the length and by the coefficient of expansion for steel, as follows:

$$\Delta L_4 = L\beta\Delta t$$

where

β = coefficient of expansion
β = 0.0000069 in./in./° F for steel.

Return again to the example problem (fig. 8.9). An evaluation of temperature effects is made as follows:

Given the following additional data:

For initial conditions:

Surface temperature = 74° F
Bottomhole temperature = 290° F @ 14,400 feet
Average initial temperature, t_i,

$$= \frac{74 + 290}{2} = 182° \text{ F}$$

For acidizing conditions:

Surface temperature = 70° F
Temperature @ packer = 90° F (Temperature cooled down from 290° F to 90° F by pumping cool fluid)
Average final temperature, t_f,

$$= \frac{70 + 90}{2} = 80° \text{ F}$$

$$\Delta t = t_f - t_i = 80° - 182° = -102° \text{ F}$$

$$\Delta L_4 = L\beta\Delta t$$

$$\Delta L_4 = 12 \times 14{,}400 \times 0.0000069 \times (-102)$$

$= -121.6$ inches.

The tubing will move upward 121.6 inches (or about 10 feet) due to the temperature effect.

Anchored vs. Unanchored Tubing

The discussion of the four tubing effects assumed that the tubing was free to move. This condition is met with a permanent packer without a locator sub or latch sub. Changes in pressure or temperature due to production or well treatment result in a change in tubing length.

If the tubing is latched into a permanent packer or if the packer is retrievable, the tubing is not free to move relative to the packer. The result is that the attempted tubing movements result in a force being imposed on the packer body. To illustrate, refer again to the example problem (fig. 8.9). The following tubing movements were calculated:

$$
\begin{aligned}
\text{Piston} &= \Delta L_1 = -111.6 \text{ inches}\\
\text{Helical buckling} &= \Delta L_2 = -\ \ 33.1 \text{ inches}\\
\text{Ballooning} &= \Delta L_3 = -\ \ 24.3 \text{ inches}\\
\text{Temperature} &= \Delta L_4 = \underline{-121.6} \text{ inches}
\end{aligned}
$$

$$\text{Total } \Delta L = -290.6 \text{ inches} = -24.2 \text{ feet.}$$

The tubing attempts to move upward, but it can't if it is latched into a permanent packer. The result is that the tubing is placed in tension. The tension force placed on the tubing can be calculated by using Hooke's law. Recall that Hooke's law is as follows:

$$\Delta L = \frac{FL}{EA_s}$$

$$F = \frac{\Delta LEA_s}{L}$$

Although helical buckling may cause unanchored tubing to shorten, it exerts only a negligible force on the packer if the tubing is anchored. In determining the amount of tension in the tubing, the helical buckling effect can be ignored.

If the seal sub is anchored to the packer body, the piston effect can also be neglected. Latching the seal subs into the packer body causes the piston force to be absorbed by the packer body. The piston force cannot push up on the bottom of the tubing string and attempt to shorten it.

The tension imparted to the tubing in the example problem, if the tubing is anchored, will be due only to temperature and ballooning effects.

$$\Delta L_{total} = \Delta L_3 + \Delta L_4$$

$$\Delta L_{total} = -24.3 + (-121.6) = -145.9 \text{ inches}$$

$$F = \frac{\Delta LEA_s}{L} = \frac{145.9 \times 30 \times 10^6 \times 1.812}{14{,}400 \times 12}$$

$$= 45{,}900 \text{ lb.}$$

Since the tubing weighs 6.5 lb/ft, the total tension load at the surface will be–

$$\text{Total tension} = 45{,}900 + 14{,}400 \times 6.5$$

$$= 139{,}500 \text{ lb.}$$

Since N-80 6.5-lb/ft tubing has a yield strength of only 145,000 lb, it would probably not be advisable to acidize the well, under the conditions outlined, with the tubing anchored. If the well were acidized without the tubing anchored, it would be necessary to have more than 24 feet of seal subs to prevent a loss of packer seal. In addition, the tubing might become permanently corkscrewed by the helical buckling effect. The job could be performed with the tubing anchored by increasing the pressure on the casing annulus or by heating up the injected fluids.

The upper left-hand portion of figure 8.12 lists the equations that were just used to calculate tubing length changes in the example problem. On the upper right-hand side of the figure, equations are given showing the force changes that the change in lengths will induce. The force equations were obtained by substituting the length-change equations into Hooke's law. The following example for temperature effects illustrates how this was done.

$$\Delta L_4 = L\beta\Delta t$$

$$F = \frac{\Delta LEA_s}{L} \text{ (Hooke's law)}$$

$$F_4 = \frac{L\beta\Delta tEA_s}{L}$$

$$F_4 = \frac{L \times 0.0000069\ \Delta t \times 30 \times 10^6}{L} \times A_s$$

$$F_4 = 207\ A_s.$$

LENGTH AND FORCE CHANGES IN TUBING

LENGTH CHANGES (all in inches)

1. Piston effect

$$\Delta L_1 = \frac{-L}{EA_s}\left[(A_p - A_i)\,\Delta p_i - (A_p - A_o)\,\Delta p_o\right]$$

2. Pressure buckling effect
(Only if Δp_i is greater than Δp_o)

$$\Delta L_2 = \frac{-r^2A_p^2(\Delta p_i - \Delta p_o)^2}{8EI(W_s + W_i - W_o)}$$

3. Ballooning effect

$$\Delta L_3 = \frac{-2L\gamma}{E}\left[\frac{\Delta p_{ia} - R^2\Delta p_{oa}}{R^2 - 1}\right]$$

4. Temperature effect

$$\Delta L_4 = L\beta\Delta t$$

FORCE CHANGES (all in pounds)

1. Piston effect

$$F_1 = (A_p - A_i)\,\Delta p_i - (A_p - A_o)\,\Delta p_o$$

2. Buckling effect (This effect can shorten tubing, but can exert only a negligible force.)

3. Ballooning force

$$F_3 = 0.6(\Delta p_{ia}A_i - \Delta p_{oa}A_o)$$

4. Temperature effect

$$F_4 = 207\,A_s\Delta t$$

TOTAL TENSION

$$\Delta L^1 = \frac{LF}{EA_s}$$

TOTAL SLACKOFF

$$\Delta L^1 = \frac{LF}{EA_s} + \frac{r^2F^2}{8EI(W_s + W_i - W_o)}$$

LENGTH AND FORCE CHANGES IN TUBING

TERMS

L = depth, in inches

E = modulus of elasticity, in psi (30,000,000 for steel)

A_s = cross-sectional area of the tubing wall,* in sq in.

A_p = area of packer ID, in sq in.

A_i = area of tubing ID, in sq in.*

A_o = area of tubing OD, in sq in.*

Δp_i = change in tubing pressure at packer, in psi

Δp_o = change in annulus pressure at packer, in psi

Δp_{ia} = change in average tubing pressure, in psi

Δp_{oa} = change in average annulus pressure, in psi

Δt = change in average tubing temperature, (°F)

r = radial clearance between tubing OD and casing ID, in inches $(ID_c - OD_t)/2$

I = moment of inertia of tubing about its diameter $I = \pi/64(D^4 - d^4)$, in inches^4 where D is OD and d is ID*

W_s = weight of tubing per inch, lb/in.*

W_i = weight of fluid in tubing, lb/in.*

W_o = weight of displaced fluid, lb/in.*

R = ratio of tubing OD to ID*

β = coefficient of thermal expansion (.0000069 in./in./°F for steel)

γ = Poisson's ratio (0.3 for steel)

*Given in chart for common sizes and weights

For extreme conditions, basic formulas and derivations, see "Helical Buckling of Tubing Sealed in Packers," by Arthur Lubinski, W. H. Althouse, Jr., and J. L. Logan, *Journal of Petroleum Technology.*

Figure 8.12. Length and force changes in tubing (Courtesy of Baker Packers)

Packer-Tubing Calculations

It should be obvious that calculations should be made before packers are run if treating operations are to be performed. Retrievable weight-set and permanent packers pose different types of problems, so they will be considered separately.

Retrievable Weight-Set Packer Calculations

It was pointed out in the discussion of forces acting on the packer body that the biggest hazard with weight-set packers in treating operations is that they might be pumped up the hole. Simple force-balance calculations that are usually made to evaluate this system were given in the discussion of figure 8.6.

In addition to the force-balance calculations on the packer body, it is advisable to consider the forces on the tubing due to ballooning and temperature effects. The increased pressure on the tubing string during treatment will cause the tubing to balloon. The ballooning effect will cause the tubing string to shorten. Since it is screwed into the packer body, it cannot contract without unseating the packer. The ballooning force should be considered in the evaluation. If cold fluids are pumped into the well, they will cause the tubing to shrink and will add an additional force on the packer.

The effects of the temperature and ballooning forces can be illustrated by the use of the example problem in figure 8.6, with additional conditions given.

Initial conditions:

Surface temperature = 74° F

Bottomhole temperature @ packer = 190° F

$$\text{Average temperature} = \frac{74 + 190}{2} = 132° \text{ F}$$

Treating conditions:

Surface treating temperature = 70° F

Bottomhole treating temperature = 85° F

$$\text{Average temperature} = \frac{70 + 85}{2} = 77.5° \text{ F}$$

$$\Delta t = t_f - t_i = 77.5 - 132.0 = -54.5° \text{ F}$$

Initial conditions:

$$p_i = p_o = 3{,}042 \text{ psi}$$

$$p_{ia} = p_{oa} = \frac{0 + 3{,}042}{2} = 1{,}521 \text{ psi}$$

Treating conditions:

$$p_i = 3{,}332 \text{ psi}$$

$$p_o = 3{,}042 + 583 = 3{,}625 \text{ psi}$$

$$p_{ia} = \frac{1{,}000 + 3{,}332}{2} = 2{,}166 \text{ psi}$$

$$p_{oa} = \frac{583 + 3{,}625}{2} = 2{,}104 \text{ psi}$$

$$\Delta p_{ia} = 2{,}166 - 1{,}521 = 645 \text{ psi}$$

$$\Delta p_{oa} = 2{,}104 - 1{,}521 = 583 \text{ psi}$$

Ballooning force, F_3:

$$F_3 = 0.6\,(\Delta p_{ia} A_i - \Delta p_{oa} A_o)$$

$$F_3 = 0.6\,(645 \times 3.13 - 583 \times 4.43)$$

$$F_3 = -\ 340 \text{ lb}$$

The ballooning effect is trying to contract the tubing, since $\Delta p_{oa} < \Delta p_{ia}$, 583 psi < 645 psi.

Temperature force, F_4:

$$A_s = 4.43 - 3.13 = 1.304 \text{ sq in.}$$

$$F_4 = 207 \Delta t A_s = 207 \times (-54.5) \times 1.304 = -14{,}700 \text{ lb}$$

$$F_3 + F_4 = -14{,}700 + (-340) = -15{,}040 \text{ lb.}$$

The temperature effect is trying to contract the tubing, since $\Delta t = -54.5°$ F.

The addition of the upward force of 15,040 lb due to the temperature and ballooning effects will cause the packer in the example problem to unseat, even though 583 psi is applied to the casing.

Downward forces:

Tubing weight on packer	= 7,000
Annular force (583 psi + 3,042 psi) × 14.81 sq in.	= 53,680
	60,680 lb

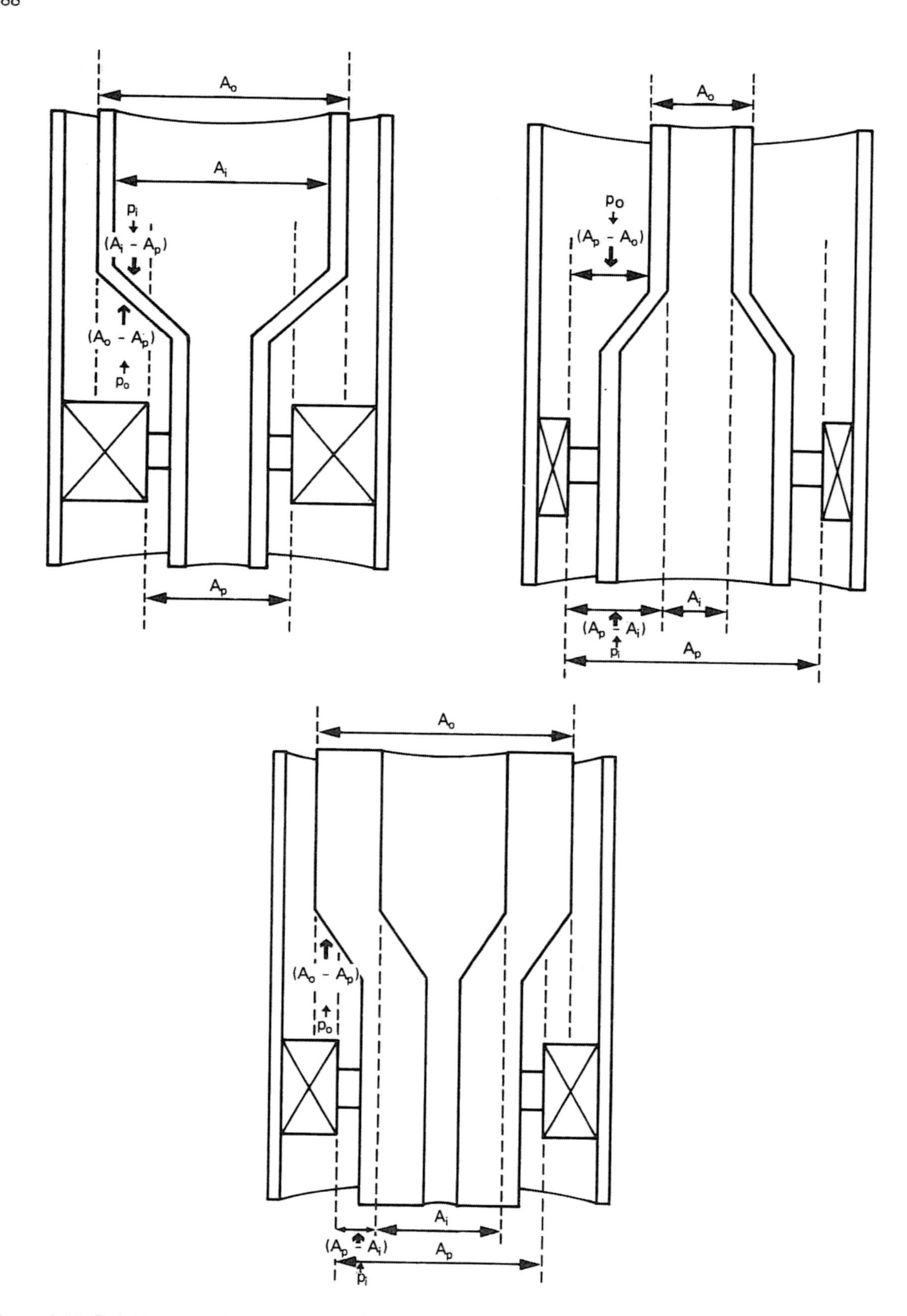

Figure 8.13. Relationship of packer seal bore area and outside area of tubing

Upward forces:

Tubing pressure (3,332 psi × 16.11 sq in.)	= 53,680
Temperature effect	= 14,700
Ballooning effect	= 340
	68,720 lb

Resultant upward force = 68,720 − 60,680
= 8,040 lb.

Temperature forces should always be evaluated. Ballooning forces are usually small and normally can be ignored.

Permanent Packer Calculations

It has been pointed out that it is normally not necessary to evaluate the forces acting on the permanent packer body. The permanent packer is so strong that the casing will usually fail before the packer body will fail.

The main concern with permanent packers then is tubing movement. If the tubing is not latched into the packer, a sufficient length of seal subs must be run to ensure that the seal subs do not pull out of the packer body and cause the pressure seal between the tubing and the casing annulus to be lost. All four of the tubing effects should be evaluated.

The piston effect is greatly influenced by the relationship between the packer seal bore area A_p and the outside area of the tubing A_o. Figure 8.13*A* depicts a situation in which the tubing outside diameter is greater than the packer seal bore diameter. The annulus pressure at the packer, p_o, acts upward on the area $A_o - A_p$. The direction of this force is upward and will tend to make the tubing contract. If the diameter of the packer seal bore area is greater than the outside diameter of the tubing (fig. 8.13*B*), the annulus pressure at the packer will act downward on the area $A_p - A_o$. The direction of the annular pressure force is now downward and will cause the tubing to elongate.

Figure 8.13*C* depicts the special case in which the seal bore diameter is between the outside and the inside diameter of the tubing. The annulus pressure imparts an upward force on the area $A_o - A_p$.

The tubing pressure (fig. 8.13*A*) acts downward on the area $A_i - A_p$. It imparts a force downward that tends to keep the seals seated. The tubing pressure (fig. 8.13*B*) acts upward on the area $A_p - A_i$ and tends to pull the seals out of the packer. The annulus pressure (fig. 8.13*C*) acts upward on the area $A_o - A_p$, and the tubing pressure also acts upward on the area $A_p - A_i$. Both of these forces will tend to pump the seals out of the packer.

The above relationships should be kept in mind when selecting packer bores for deep high-pressure installations using heavy fluids. An overriding concern to have as large an opening as possible will influence the decision. If such a problem does not exist, then the size of the seal bore can be selected to minimize tubing movement.

Temperature Profiles in Tubing

It is necessary to estimate tubing temperature profiles for packer calculations if measured data are not available. The following discussion of typical temperature profiles gives guidelines for making reasonable estimates.

First consider the temperature distribution when the well is shut in (fig. 8.14). The ambient

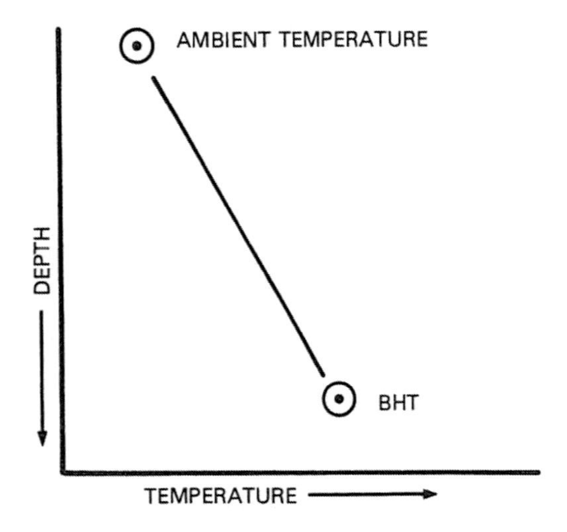

Figure 8.14. Typical shut-in temperature profile

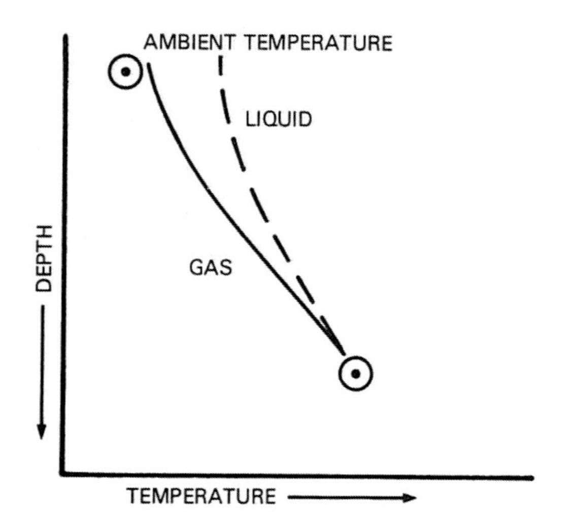

Figure 8.15. Flowing temperature gradient

surface temperature can be estimated at 70° – 74° F in the U. S. The bottomhole temperature (BHT) can be estimated by using thermal gradient data from nearby wells or from logging companies. Local logging companies can usually supply the correct thermal gradient to use, as well as the average ambient temperature. The shut-in temperature gradient is a straight line connecting the surface and bottomhole temperatures.

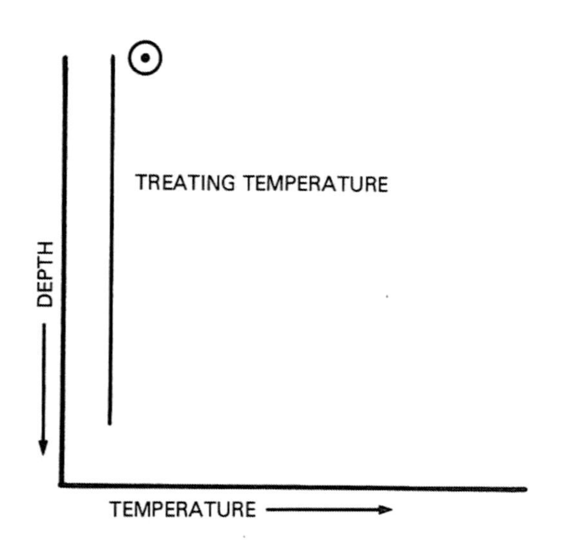

Figure 8.16. Treating temperature profile

When the well starts producing, a temperature profile similar to the one shown in figure 8.15 develops. The bottomhole temperature is unchanged from the shut-in condition, but the flowing surface temperature increases. The amount of increase will depend on the flow rate. Surface flowing temperatures with liquids flowing will increase more than those with gas flowing. When the well is treated by pumping fluids into the well, a profile similar to the one shown in figure 8.16 results. The treating temperature is shown to be lower than the average ambient temperature, and it is unchanged going downward into the hole. This is the condition that applies when cold fluids are pumped in the winter months for several hours.

Generally the fluid picks up little heat as it moves down the hole, although it obviously picks up some and is not a straight line as shown. If actual data are lacking and pumping operations are to be continued for several hours, an estimate similar to the one shown will be conservative.

During the winter months treating fluids become cold, and severe contraction of the tubing can result. Sometimes it is necessary to heat up the fluids before injecting them, if calculations indicate that tubing contraction will be enough to pull seal subs out of a permanent packer.

Prevention of Buckling

Helical Buckling

Helical buckling occurs when the pressure (p_i) at the packer in the tubing is greater than the annulus pressure (p_o) at the packer. If $p_o > p_i$, then buckling will not occur. This information is useful. If helical (pressure) buckling occurs, it will not be possible to run wireline tools to bottom because of the corkscrewing of the tubing. Sometimes this problem can be solved by pressuring up on the casing annulus. The pressure placed on the annulus should not exceed the safe working pressure of the casing. If the annulus pressure can be increased enough to reduce the buckling, the wireline tools will go on to bottom.

Mechanical Buckling

Mechanical buckling is corkscrewing of the tubing caused by slacking off too much tubing weight in setting packers. Many people assume that any tubing weight slacked off at the surface will automatically be transmitted to the packer, but this is not always the case, especially as larger weights are slacked off. As the slack-off is started, the tubing tends to corkscrew and rub against the casing wall. The friction between tubing and casing prevents all of the weight slacked off at the surface from reaching the packer. As more weight is slacked off, the corkscrewing effect becomes more pronounced, and progressively less slacked-off weight reaches the packer.

Most packer company handbooks have charts to estimate the amount of slack-off weight that reaches the packer. The total slack-off equation given in figure 8.12 can be used to calculate this force. The solution of this equation for force can be time-consuming, so charts should be used if they are available.

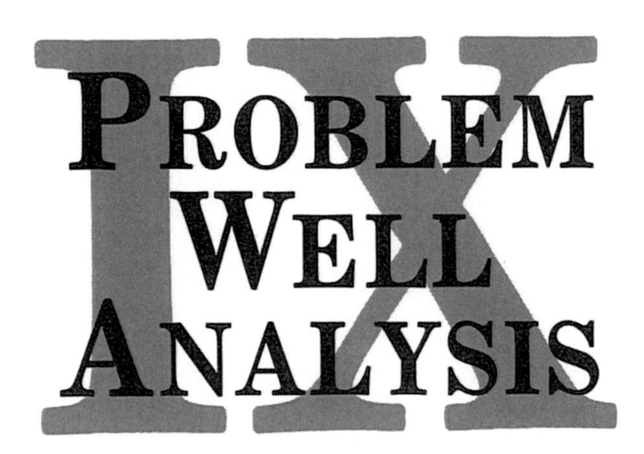

What Problem Well Analysis Is

Problem well analysis is one of the most important phases of production operations. Its purpose is to determine whether a well has a problem, then to identify the problem and thus be in a position to recommend a workover to correct it. The ultimate outcome of problem well analysis is often a workover.

Since workovers can be very expensive, it is imperative that the well be thoroughly analyzed before workover operations begin. Problem well analysis is the same as the diagnostic procedure used by a doctor. The workover is comparable to surgery, which is performed only after careful diagnosis.

If a well suddenly goes off production, there is no doubt that some type of problem exists. If it is a pumping well, the odds are that the problem is mechanical. If the well produces from an unconsolidated sand, it is possible that the well has sanded up. In any event, the well data must be carefully reviewed to determine the actual problem.

A well can have a problem without exhibiting any obvious symptoms. Two wells may be producing at the same rates, yet one may have skin damage that restricts its production. The skin damage can be determined by running and analyzing a pressure buildup test.

Problem wells are identified by determining whether their behavior deviates from normal production behavior. Problem well analysis involves the determination of abnormal or anomalous behavior. In order to identify abnormal or anomalous behavior, it is necessary to know what constitutes normal behavior. Reservoir characteristics and drive mechanisms are important to know. The expected performances for dissolved-gas drive, gas-cap drive, and water drive wells vary considerably; such information is needed to identify and analyze problem wells. For example, if a well starts to make water, its behavior is anomalous if it is producing from a dissolved-gas drive reservoir. On the other hand, if the well is producing from a water drive reservoir, its performance is probably normal.

Well Analysis Tools

A number of "tools" are used in analyzing well performance. The most commonly used ones are (1) well performance curves, (2) well status maps, (3) well histories, (4) wellbore sketches, (5) bottomhole pressure data, (6) fluid analyses, (7) fluid levels, and (8) others.

Well Performance Curves

Periodic well tests offer the most readily available data for problem well analysis. In order to be easily analyzed, they are usually plotted versus time on graph paper. Since the type of data available for oilwells is different from that available for gas wells, oil and gas well performance curves will be discussed separately.

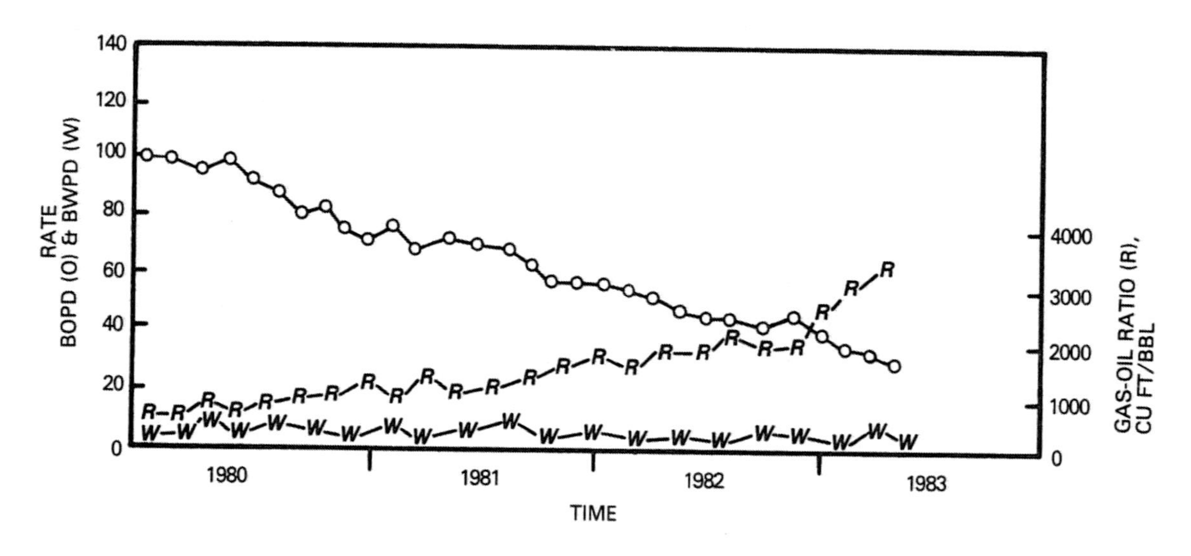

Figure 9.1. Performance curve of dissolved-gas drive oilwell

For oilwells. The performance shown in figure 9.1 is typical for an oilwell producing from a dissolved-gas drive reservoir. Oil and water rates, as well as gas-oil ratio (GOR), are plotted vs. time.

The oil production rate that is steadily decreasing is typical of a dissolved-gas drive reservoir. The gas-oil ratio steadily increases until the first part of 1983, when a more rapid increase occurs. Such change is also typical of dissolved-gas drive performance. The well is making a small amount of water, but note that

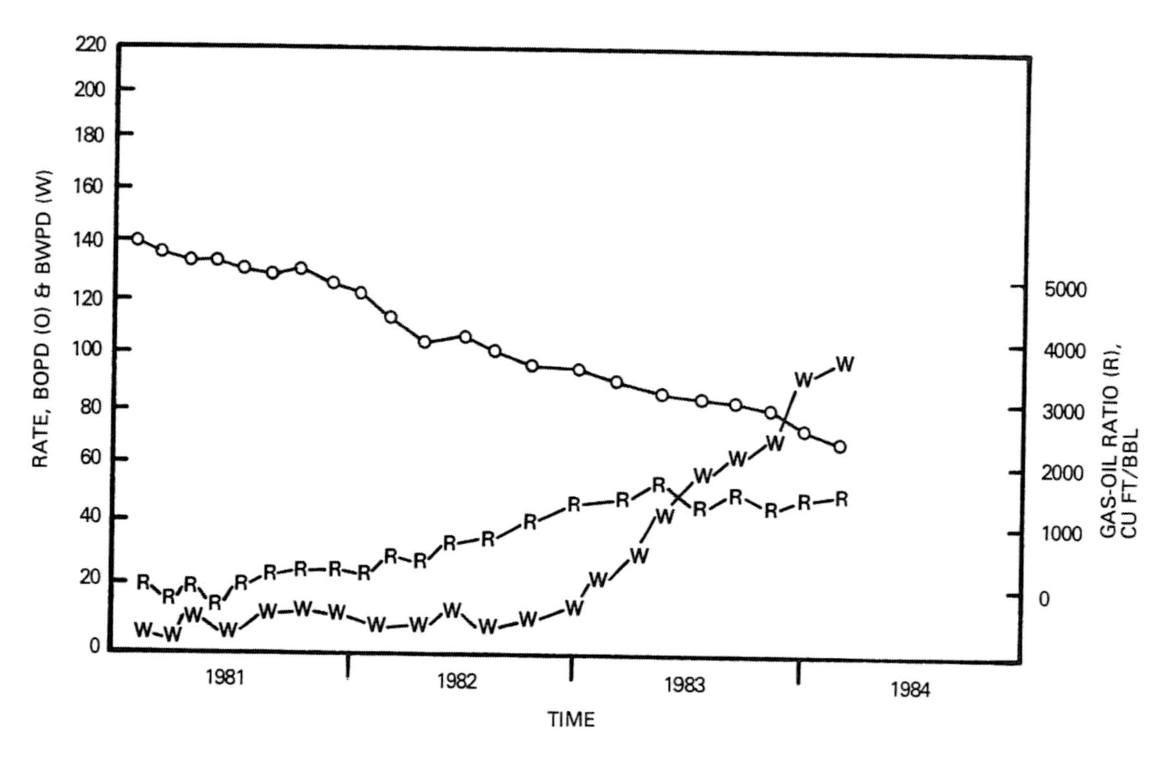

Figure 9.2. Dissolved-gas drive oilwell with extraneous water production

the water production rate remains constant. It is not unusual for a dissolved-gas drive well to make water, but it should be of a low volume and should not increase. Normal performance for a dissolved-gas drive oilwell is represented by the figure.

Figure 9.2 also concerns a well producing from a dissolved-gas drive reservoir. Note that the oil production rate steadily declines as expected. The gas-oil ratio trend is generally upward, typical for dissolved-gas drive performance. The water production rate is low and unchanging until early 1983, when the water producing rate starts to increase rapidly, and water production soon exceeds oil production. This behavior is obviously anomalous for a dissolved-gas drive well and indicates that extraneous water is apparently being produced.

Figure 9.3 represents normal performance for an oilwell producing from a water drive reservoir. Note that the producing rate is relatively constant, and that the GOR is unchanging until the early part of 1983. In early 1983 a sharp increase in the water production and a corresponding decrease in the oil producing rate occurs. This crisscrossing of the oil and water rate curves is typical of a water drive well approaching depletion. The water has broken through to the wellbore, so the water rate will continue to increase and the oil production rate will drop accordingly. This type of well is normally produced until the water-oil ratio (WOR) becomes too high to be economical.

Figure 9.4 depicts an oilwell with a mechanical problem. Note that the oil production rate declines steadily until mid-1982, when the GOR starts to increase, indicating that this well is producing from a dissolved-gas drive reservoir. The water production is relatively high, but it is unchanging, so a dissolved-gas drive would still be inferred. In the early part of 1983 both the oil and the water production rates drop rapidly to near zero, indicating that a mechanical problem, possibly a pump failure, has occurred.

For gas wells. Gas wells can be analyzed by performance curves in a manner similar to that for oilwells. In addition to well test performance plots, another useful plot for gas wells is the *P/Z* vs. cumulative production plot (fig. 9.5). The plot applies only to gas wells producing from closed or volumetric reservoirs. It does not

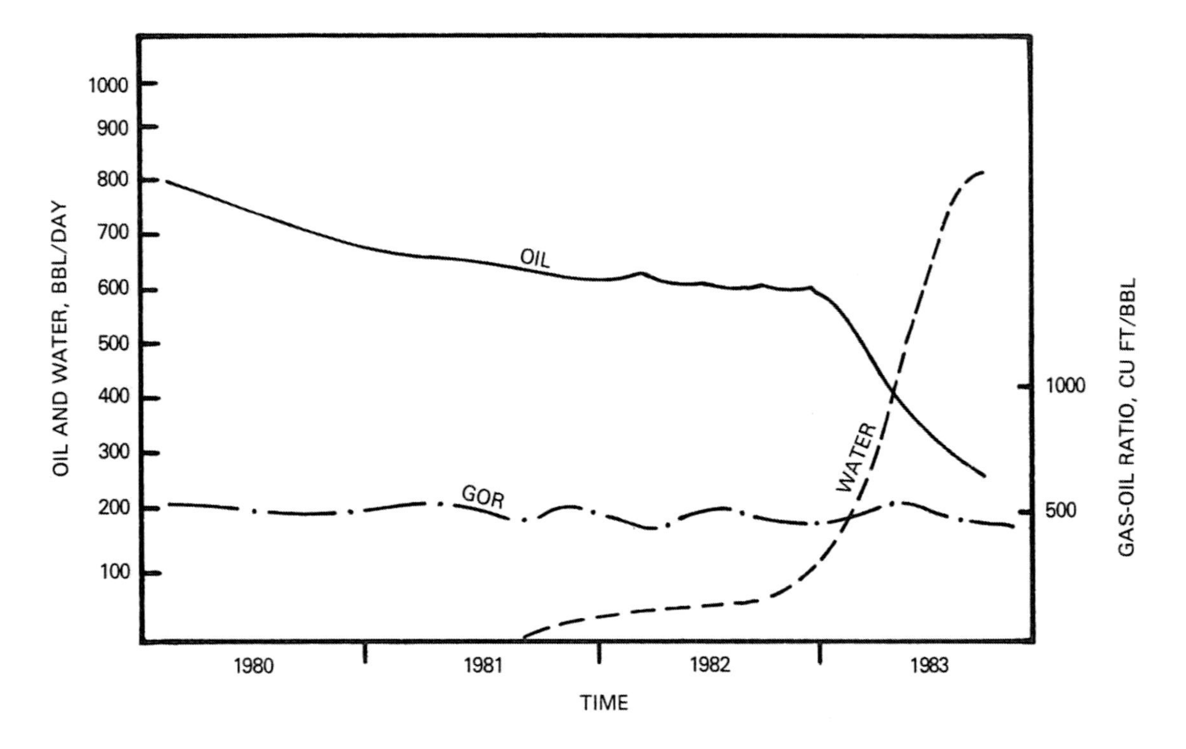

Figure 9.3. Water drive oilwell

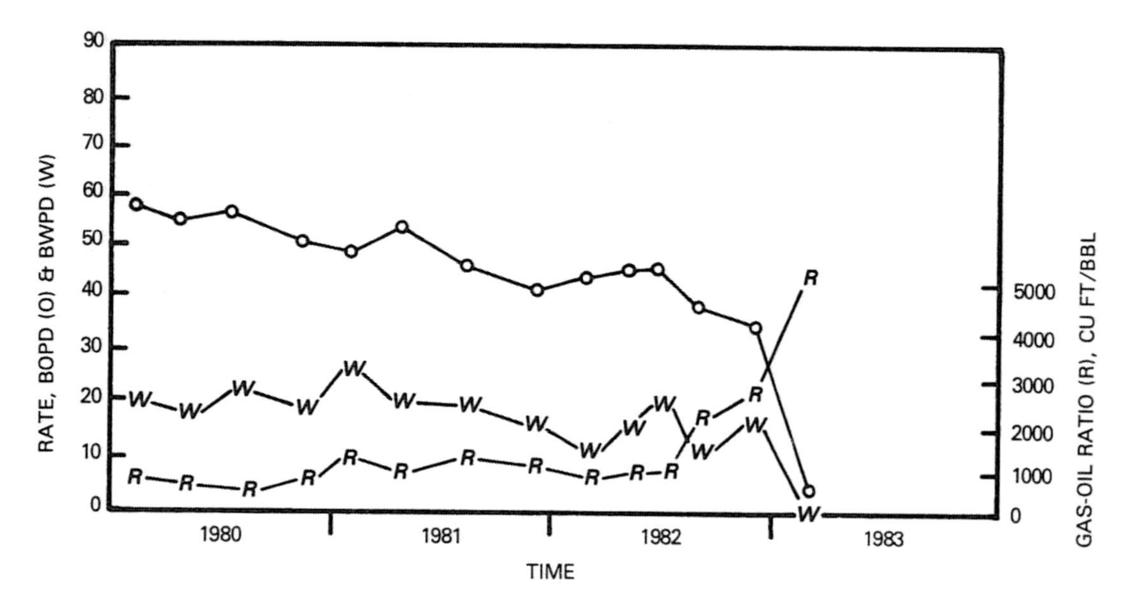

Figure 9.4. Oilwell with mechanical problem

apply to water drive gas wells. The shut-in reservoir pressure divided by the Z factor is plotted vs. the cumulative production at the time the pressure is taken. The Z factor determines how much the gas varies from the ideal gas laws. It is used to get a straight-line relationship. The curve can be thought of as P vs. cumulative production.

If the P/Z curve is extrapolated to $P/Z = 0$, the intercept with the X-axis will indicate the gas that is in place for the well's drainage area. The recoverable reserves can be estimated by extrapolating the curve to the estimated abandonment pressure Z. The P/Z curve is used to estimate gas reserves by reservoir engineers.

The P/Z vs. cumulative production curve can be made more useful by adding the rate (q) and the flowing tubing pressure (FTP) to the plot (fig. 9.6). The FTP and q plots vs. cumulative production will also *indicate* the general order of magnitude of the reserves for the well. These two plots are not as valid technically as the P/Z, but will indicate an approximate value of reserves. If the well has wellbore damage, the FTP and q plots will indicate substantially fewer reserves than the P/Z plot. They provide a very good way to determine whether skin or wellbore damage has occurred. A pressure buildup test is normally run to confirm the presence of skin damage.

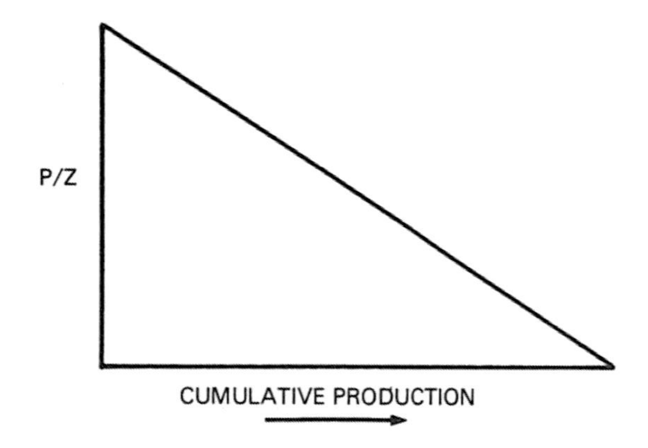

Figure 9.5. P/Z vs. cumulative production for gas well

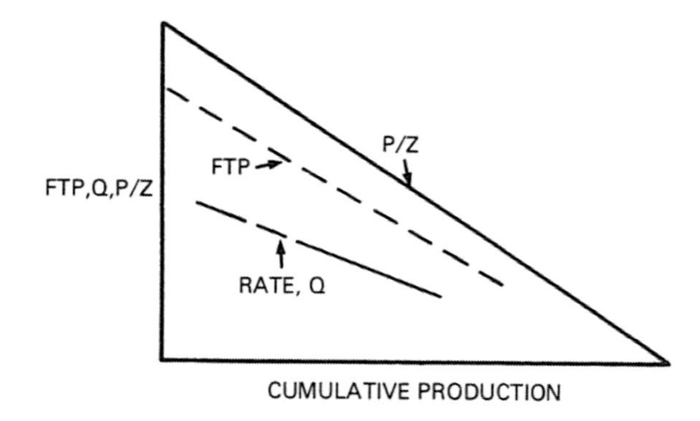

Figure 9.6. Rate and flowing tubing pressure added to P/Z vs. cumulative production plot

Well Status Maps

The well status map is one of the most useful tools available for problem well analysis (fig. 9.7). The example well status map is for a field under waterflood, so it is necessary to explain a bit about waterflooding. The producing oilwells are indicated by the circles. The injection wells, where water is injected into the producing formation, are marked with arrows. The waterflood pattern shown is called a *five-spot.* This means that each producing well is offset by four injection wells, making a five-spot pattern. A five-spot waterflood, such as the one depicted above, is usually performed on a dissolved-gas drive field. The primary recovery from a dissolved-gas drive field is low, and a waterflood is often used to increase oil recovery. Typically the oil recovery from a dissolved-gas drive reservoir is doubled by waterflooding.

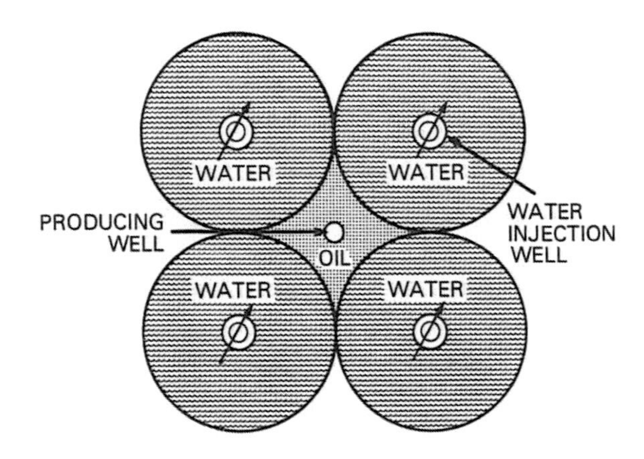

Figure 9.8. Typical five-spot waterflood pattern

In a typical five-spot waterflood (fig. 9.8), water is injected into the four injection wells at the corners of the five-spot. In the ideal case, the water moves radially from each injection well

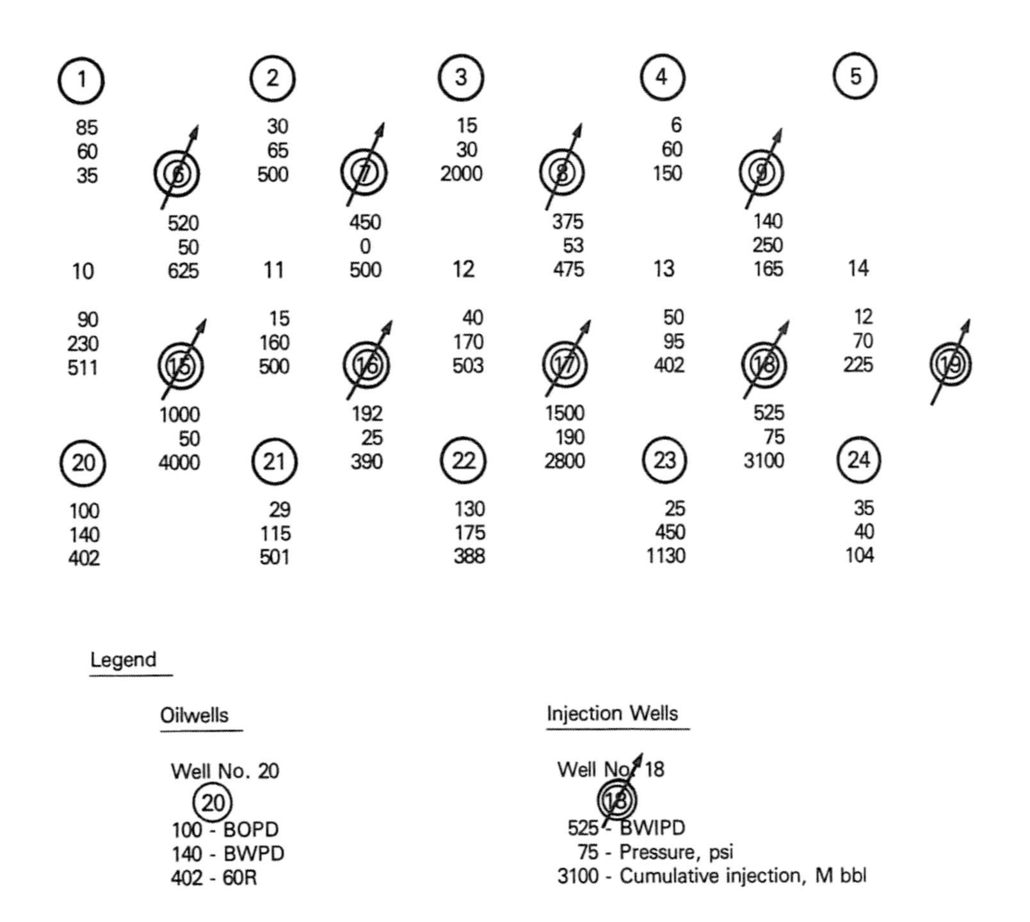

Figure 9.7. Well status map

and displaces the oil to the producing well in the center of the five-spot, as shown. There are other types of waterflood patterns, but the five-spot is typical and illustrates how many waterfloods are operated.

Referring again to the well status map (fig. 9.7), note that the latest well test is plotted below each producing oilwell. The oil rate, water producing rate, and GOR are listed for each oilwell. The injection rate in barrels of water injected per day (BWIPD), the surface injection pressure, and the cumulative water injected are listed below each water injection well.

The well status map permits a quick look at a field to evaluate the comparative performance of all the wells. It will then be apparent when one well is performing better or worse than the nearby offsets. For example, look at injection well no. 17. It is injecting 1,500 BWPD at a surface pressure of 190 psi. The offset well to the east, no. 15, is injecting only 525 BWPD at 75 psi. An actual problem may or may not be indicated, but this anomaly should be checked out further. The same procedure can be used to evaluate the oilwells and see how any well is performing as compared to the nearest offsets.

Well Histories

A well history is the history of all operations that have been conducted on a well to date. The history is usually prepared in a tabular form as shown below.

The well history can be reviewed to see whether the well has any recurrent problems. For example, if several entries are noted regarding the well sanding up, it would be desirable to check to see if any subsequent trouble is caused by sand production.

Wellbore Sketches

The wellbore sketch (fig. 9.9) supplements the well history and contains some of the same

Well History

Well Name:		Rita Santa No. 1
Location:		660' FNL and 660' FEL of Lease
Completed:		7-6-74
Elevation:		265'
Initial Potential:		Flowed 220 BOPD, 0 BWPD, GOR 672 CFB, FTP 600 psi
Casing:		8⅝" ODCSA* 1,600' w/600 sx* 5½" ODCSA 5,300' w/650 sx
TD:		5,300'
Producing Intervals Present:		A-1, A-2, and A-4 sands
Initial Completion:		Perf A-4 Sand, 5,240'–48' w/4 JSPF
Workovers:	3-13-76	Squeezed perfs 5,240'–48' w/75 sx Well producing 4 BOPD, 650 BWPD on gas lift before squeeze.
	3-17-76	Perf A-2 Sand 5,065'–77' w/4 JSPF Well flowed and swabbed 60 BOPD
	3-19-76	Acidized w/1,000 gal mud acid. Well flowed 140 BOPD w/650 GOR
Current Production:		40 BOPD, 2 BWPD – Pumping

**ODCSA:* outside diameter casing set at; *sx:* sacks

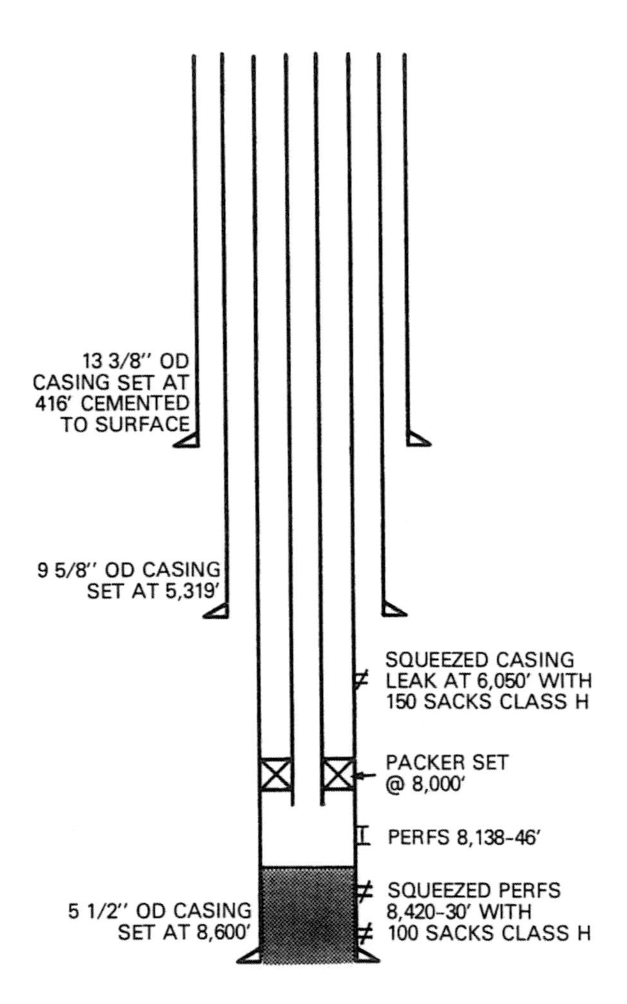

Figure 9.9. Wellbore sketch

information. It shows at a glance what equipment is in the well. This information is valuable in planning workover operations as well as being of help in diagnosing well problems.

Bottomhole Pressure Data

Static bottomhole pressures, taken by shutting the well in for a specified number of hours, are useful in evaluating well performance. In a plot of the static pressure versus time, the trend of the pressures will indicate the type of drive mechanism. In waterfloods, the pressure data will help to identify waterflood response.

Bottomhole pressure buildup (or fall-off) tests are obtained by shutting in a well and obtaining the bottomhole pressure reading versus time. Buildup tests can be used to determine static bottomhole pressure, interwell *kh*, flow efficiency, and skin effect. The flow efficiency is very useful in analyzing wells for stimulation.

Flowing and static bottomhole pressure data, in conjunction with production tests, can be used to determine productivity index data (PI = BOPD/psi-drawdown). PIs can be used to compare well performance and estimate production increases from lowering back-pressure. PI is primarily applicable to water drive and gas-cap expansion drive oilwells. It can also be used to evaluate waterflood wells. Inflow performance analysis for solution-gas drive reservoirs below the bubble point should be done with Vogel's equation.

Fluid Analyses

Fluid analyses are very helpful in analyzing problem wells. Since the chemical composition of water from different formations normally varies, water analysis can be used to identify extraneous water production. Extraneous water can come either from a casing leak or from a bad cement job that causes communication behind the pipe with another zone. Water analysis can also be used to monitor waterflood performance when the injection water is different from the formation water, especially in waterfloods where fresh water is injected. It is possible to tell when the injected water breaks through into a producing well by observing the change in the analysis of the produced water.

Oil analyses are not as useful as water analyses, but sometimes it may be possible to verify communication through a packer on a multiple completion if the oil composition is sufficiently different in the various zones.

Fluid Levels

It was pointed out in the discussion of water well performance in chapter 4 that the height of the fluid column in the annulus while the well is producing reflects the producing bottomhole pressure. The same thing is true of oilwells. If the height of the fluid column in the annulus is

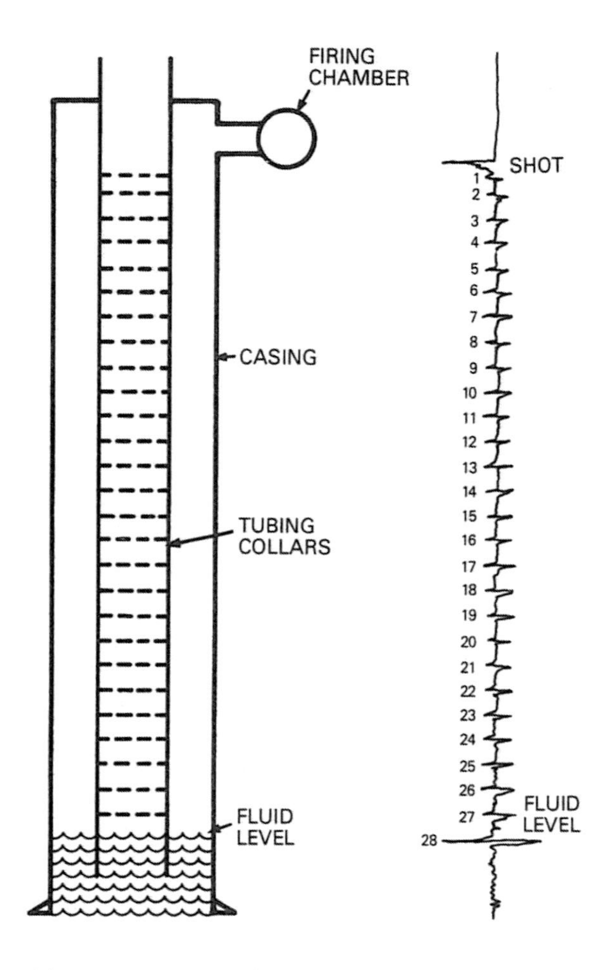

Figure 9.10. Acoustic well-sounding device

known, an estimate of the bottomhole pressure can be made. Fluid levels in oilwells are obtained by acoustic well-sounding devices (fig. 9.10).

The acoustic well-sounding device has a firing chamber that is attached to the casing annulus. A blank cartridge such as a .45 caliber is fired in this chamber. The explosion sends a sound wave down the tubing-casing annulus. As the sound wave passes each tubing collar, it is reflected back to the surface. A recording device with a strip chart records the reflections. When the sound wave hits the fluid level, it sends back a large reflection. By counting the number of joints of tubing above the fluid level, the depth to the fluid can be calculated by using the tubing tally.

Fluid-level devices can be used on flowing or pumping wells. Their main application is for pumping wells, especially low-capacity wells producing from dissolved-gas drive reservoirs. Recall from the discussion of inflow performance in chapter 4 that the maximum flow rate, q, is obtained with the largest value of drawdown, $p_e - p_w$. On low-capacity pumping wells, an effort is made to keep the value of p_w as low as possible. This is accomplished by pumping the well as close to the pump as possible and by running the pump opposite the producing formation. Ideally, a fluid-level shot on a low-capacity pumping well should show about 100 to 200 feet of fluid above the pump while the well is producing.

Other Well Analysis Tools

Many other well analysis tools such as posted logs, log cross sections, isopachous maps, and structure maps are used in problem well analysis.

Problem Well Analysis Examples

Declining Oil Production from Oilwell

Typically, a well problem is first detected by use of the well performance curve, usually available for all wells. If there are a large number of wells in a field, the well test data are usually stored in a computer file and printed out periodically, such as once every three months. Most computer programs are designed to flag, or mark, the wells that have had an unusual change in producing rates. If there is no computer program, then curves are manually reviewed to pick out anomalous performance.

Assume that the well performance curve shown in figure 9.11 is observed during the periodic review of a field. A sudden drop in the oil production rate is noted immediately. It is obvious that something is wrong with this well. If the reservoir drive mechanism is not known, it is obvious from the decreasing oil rate,

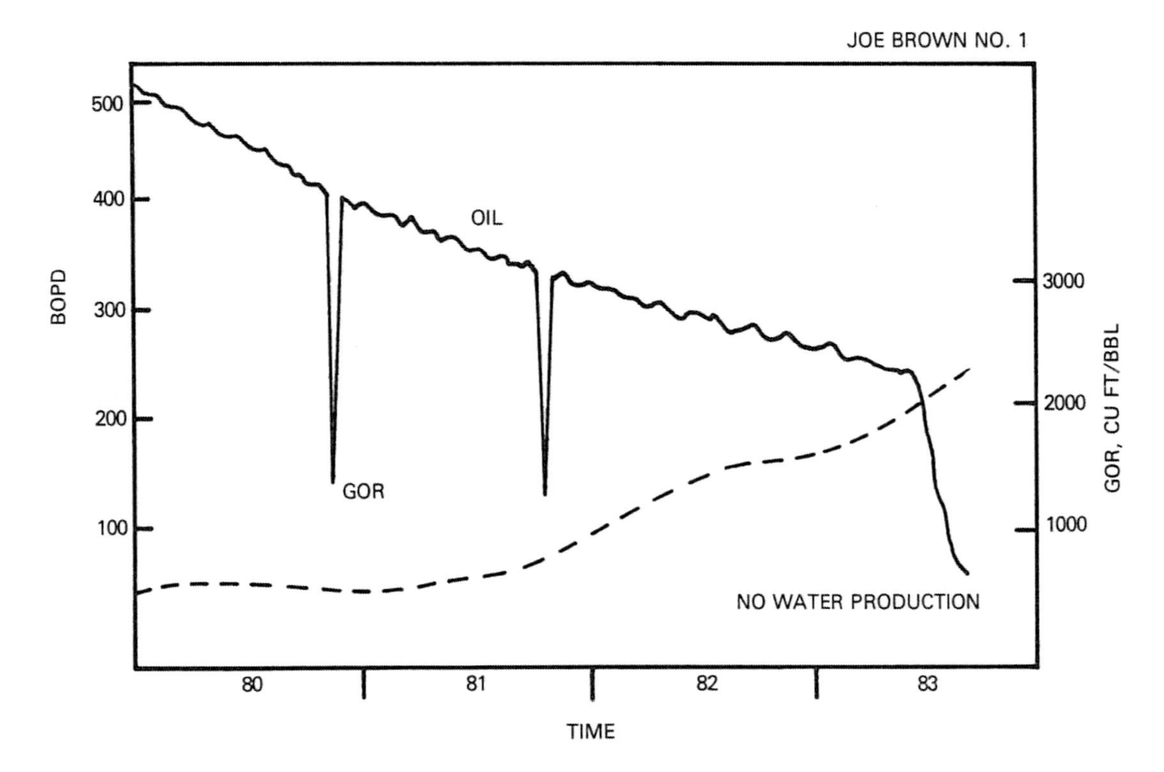

Figure 9.11. Well performance curve

increasing GOR, and lack of water production that this well probably produces from a dissolved-gas drive reservoir. The sudden drop in the oil rate obviously represents anomalous behavior.

A check of the records shows that this well is being produced by a beam pumping unit. If a well is being artificially lifted, a sudden decrease in oil production usually indicates a pumping equipment failure. This being the case, an acoustic fluid-level check is made, showing that the producing fluid level is 200 feet above the pump. This fluid-level data eliminates the pump and rods as the potential problem. Their failure would be indicated by a high fluid level.

At this point it is a good idea to look at a wellbore sketch and a well history to see if there are any other clues. Review the wellbore sketch (fig. 9.12) and the following well history.

A problem was probably spotted from reviewing the well history. The well was off production in 1980 and 1981 due to sand production. The performance curve indicates where this occurred.

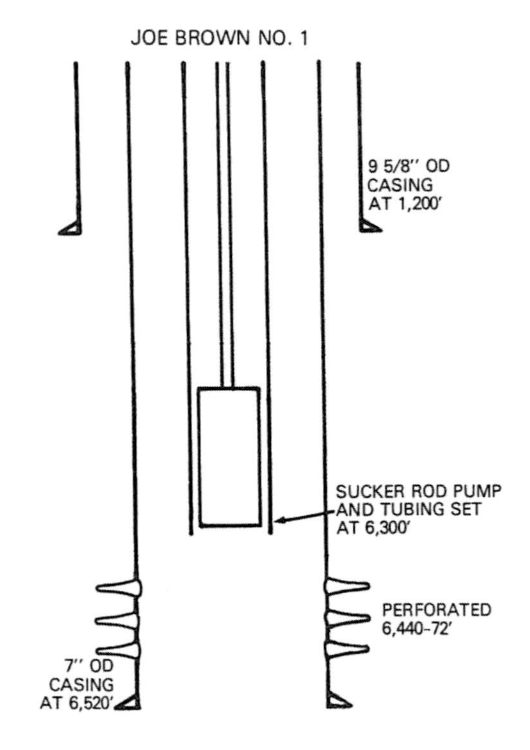

Figure 9.12. Wellbore sketch

Well History

Well Name:		Joe Brown No. 1
Location:		660' FNL and 660' FEL of lease
Completed:		March 4, 1979
Elevation:		75'
Initial Potential:		Flowed 520 BOPD, 0 BWPD, GOR 520, FTP 225 psi
Casing:		10¾" ODCSA 1,200' with 300 sacks of Class A cement with 4% bentonite 7" ODCSA 6,520' with 300 sacks of Class H cement
Total Depth:		6,520'
Producing Intervals Present:		Frio S-1 Sand, 6,440'–6,472'
Initial Completion:		Perforated the S-1 Sand 6,440'–6,472' with 4 jet shots/foot
Workovers:	2-1-80	Well off production. Pulled pump, replaced leaking traveling valve, and returned well to production. Well tested 490 BOPD, 0 BWPD.
	11-14-80	Well off production. Pulled rods and tubing. Ran steel line and found sand fill at 6,410'. Washed out sand. Returned well to production at test rate of 450 BOPD and 9 BWPD.
	10-17-81	Well off production. Pulled rods and tubing. Found pump sanded up. Reran new pump. Well tested 420 BOPD and 0 BWPD.

The well history shows that the well was completed by setting the casing through the Frio S-1 Sand and then perforating the section 6,440' – 6,472' with 4 jet shots/foot. Indications are that the present problem might be due to sand fill, since the same problem has occurred twice in the past three years. A pump problem has been eliminated by checking the fluid level, which was found to be 200 feet above the pump.

The next step is to pull the rods and tubing and run a steel line to check the total depth; a sand fill is found at 6,400 feet. The problem has now been identified, and the problem well analysis phase is ended.

Since the well has sanded up twice previously, consideration should be given to running a screen and liner. Since the well has 7" OD production casing, there is enough clearance to run a large screen and liner, which should eliminate the sand problems and still permit the well to produce at its current rate.

A well status map is not used because the problem was indicated by the anomalous performance noted on the well performance curve. A well status map would have shown that the producing rate of the Joe Brown No. 1 was considerably less than offset wells. This flag could also have triggered the study. Although the example given is admittedly a textbook one, it illustrates the procedure to follow in analyzing a problem oilwell.

Declining Gas Production from Gas Well

The following actual example of a gas well illustrates the use of the performance curve and the *P/Z* rate and FTP vs. cumulative production

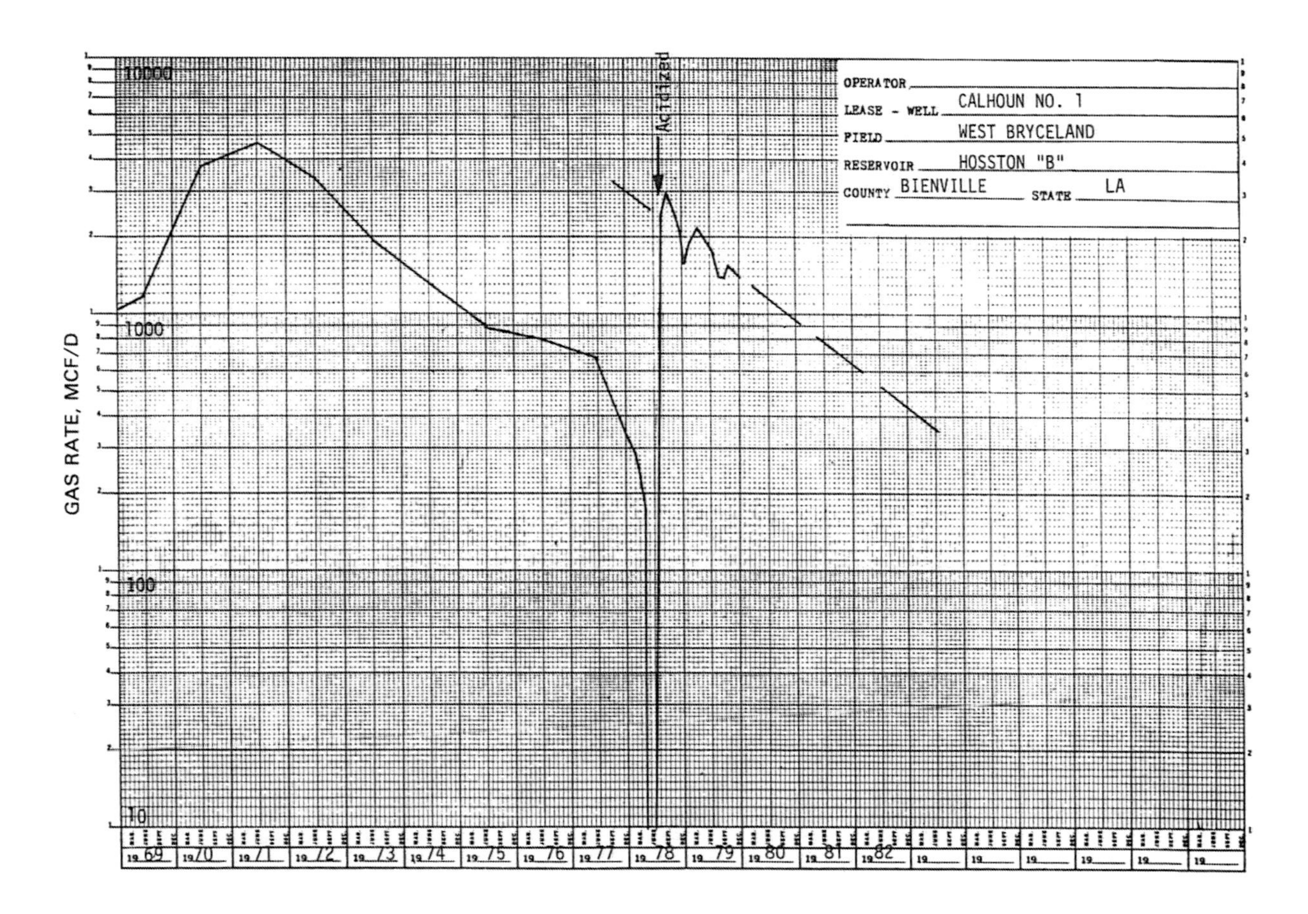

Figure 9.13. Performance curve

curves to spot a formation damage problem. A performance curve for the Calhoun No. 1 is given in figure 9.13.

Note that production declined from a high of about 4,600 Mcf/d in 1971 to about 680 Mcf/d in mid-1977. The production then dropped precipitously, and the well went off production in 1978.

The *P/Z* rate and FTP vs. cumulative production curves for the Calhoun No.1 are given in figure 9.14.

The FTP data are scattered and do not indicate a trend, but there are two trends on the rate curve. The first trend in effect prior to 1978 indicates that the well should recover about 7.4 billion cubic feet (Bcf) of gas. The *P/Z* curve indicates that the well should recover approximately 9.6 Bcf if produced to an abandonment bottomhole pressure of 700 psi. The rate curve is extrapolated to a rate equal to zero to get ultimate recovery, while the *P/Z* is extrapolated to the estimated abandonment bottomhole pressure. An estimate of an abandonment pressure of 700 psi is reasonable for this well, which is approximately 7,000 feet deep. An abandonment pressure of 100 psi/1,000 ft of depth is a good rule of thumb here.

Since the rate curve indicates an ultimate recovery 2.2 Bcf less (9.6 − 7.4 = 2.2) than the *P/Z* curve, formation damage is a good diagnosis of the problem. A pressure buildup test can be used to confirm the existence of wellbore damage. The well was acidized, and production increased to 3,000 Mcf/d, as shown on the performance curve. A second rate trend is shown on the rate vs. cumulative plot, indicating that the well will recover about 9.2 Bcf, or slightly less than was estimated by the *P/Z* vs. cumulative method. This agreement is very close.

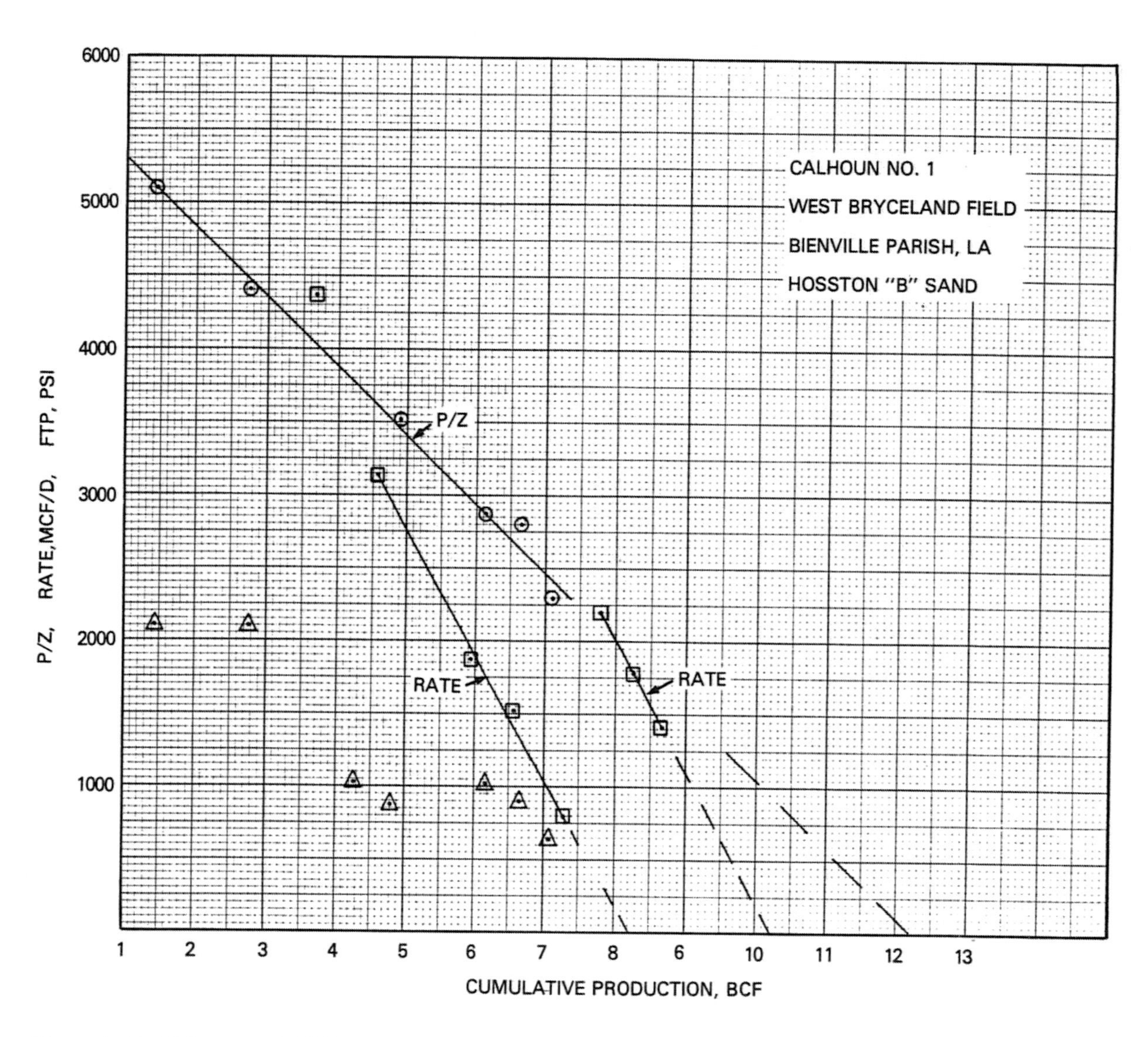

Figure 9.14. P/Z rate and FTP vs. cumulative production

X Workover Methods

Two basic types of workover methods are (1) conventional and (2) unconventional. A conventional workover generally involves killing the well and moving in a rig large enough to remove the tubing string and associated downhole equipment. The remedial work that is required, such as squeezing and perforating, is performed. The tubing and any other downhole equipment is rerun, and the well is placed on production. The key to conventional workovers is that the well is killed and the tubing string removed.

Unconventional workovers are (1) concentric-tubing workovers; (2) coiled-tubing workovers; (3) wireline workovers; and (4) pump-down, or through-flowline workovers.

In addition to the use of the two basic types of workovers, snubbing units can be used on either conventional or concentric-tubing workovers to permit the work to be done under pressure without the well's being killed.

Conventional Workovers

Well-Killing Procedures

Conventional workovers are normally performed by first killing the well. Before performing some unconventional workovers, it may also be necessary to kill the well. A well is killed by filling the hole with a fluid that will exert a pressure greater than the formation pressure. Three methods may be used to place the kill fluid in the hole.

Circulating method. The circulating method is preferred and is the one most often used. Fluid is pumped down the casing annulus and back out the tubing string, or vice versa (fig. 10.1). The well is circulated until all of the oil and gas present in the annulus is removed, and a kill fluid with a hydrostatic head greater than the shut-in bottomhole pressure is left in the well as a "load" fluid. For example, assume that a well has a bottomhole pressure of 3,120 psi and a depth of 6,500 feet. If a 100-psi safety factor is used, fluid with a gradient greater than (3,120 + 100)/6,500 = 0.495 psi/ft must be used. Since fresh water weighs 8.33 lb/gal and exerts a pressure of 0.433 psi/ft, the weight of the kill fluid can be calculated as follows:

$$\text{Kill fluid weight} = \frac{0.495}{0.433} \times 8.33 = 9.5 \text{ lb/gal.}$$

To get a kill fluid of 9.5 lb/gal, the weight of salt water can be increased by adding $CaCl_2$. A weight of about 10 lb/gal can be achieved by adding $CaCl_2$. A weight of up to 16 lb/gal can be obtained by adding bromines.

If the tubing is set on a packer, the packer must be released or a circulating valve opened to permit circulation. If the packer cannot be released and there is no circulating valve, a hole can be punched in the tubing to permit circulation.

Bullheading. If the well cannot be circulated and it is possible to pump into the formation, the well can be killed by pumping (bullheading) the kill fluid into the well. Here again a fluid with

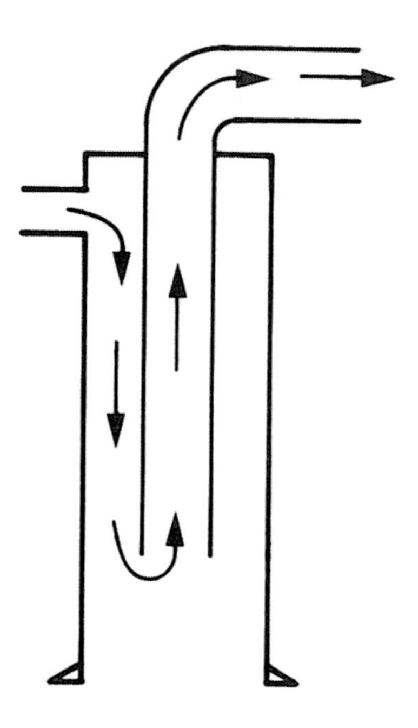

Figure 10.1. Circulation method for killing well

sufficient density is needed to overcome the bottomhole pressure and keep the well killed.

The problem with bullheading is that any junk, mill scale, pipe dope, or debris in the wellbore will be pumped into the perforations and may damage the well. Also it is difficult to pump into some low-permeability wells without exceeding the frac pressure.

Lubricating and bleeding. Sometimes the well cannot be circulated and the perforations are plugged; killing the well by bullheading is thus prevented. The lubricating and bleeding technique can then be used. A high-density fluid with enough weight to kill the well is pumped into the tubing string until the maximum allowable tubing pressure is reached. The well is then shut in for a period of time to permit gravity segregation of the fluids in the tubing. The lower-density oil or gas will come to the top of the tubing string, and the high-density kill fluid will settle down the hole. The oil or gas is bled off, reducing the pressure in the tubing. The well is shut in as soon as the high-density kill fluid starts to flow back. More kill fluid is pumped in, and the process is repeated until the hole is full of kill fluid. The procedure can be time consuming, especially if the tubing is full of oil. Since gas is more compressible than oil, it doesn't take as long to kill a well when the tubing is full of gas, since more kill fluid can be pumped in on each cycle.

After Well Killing

After the well has been killed in a conventional workover, blowout preventers are installed, and the tubing string and any other bottomhole equipment are removed. Blowout preventers are normally installed even though the well is killed. All work is performed through the preventers so that the well can be kept under control if it kicks. The only exceptions are some low-pressure pumping wells that can be safely pulled by just loading the hole. If there is any question, blowout preventers should be used.

If the workover calls for squeezing off perforations, it is easily performed by running a tubing work string back into the hole, with a packer if necessary. Drilling out cement from the squeeze is easily accomplished by running the tubing work string with a bit. In fact, any type of operation can be easily performed because the full casing ID is available to work in. The well can be perforated with a casing gun in order to get the benefit of the largest-size charges, minimum standoff distance, and 90° phasing for maximum productivity. The principal problem is that some formations can be harmed by the kill fluid. In addition, some wells have such low bottomhole pressures that if they are killed with fluid, they don't have enough pressure to make the kill fluid flow back, and the well can be lost.

Unconventional Workovers

Since conventional workovers have so many advantages, the obvious question is why use anything else. The answer is cost. Conventional workovers on offshore wells can be very costly, sometimes running over one or two million dollars. Unconventional-workover technology has been developed in order to keep costs down.

Concentric-Tubing Workovers

A concentric-tubing workover is performed like a conventional workover except that the tubing is not pulled and all of the equipment used is smaller and lighter, since it has to go down the tubing string. In a concentric-tubing workover, the blowout preventer stack is installed on top of the Christmas tree (fig. 10.2). Therefore, the well is under complete control at all times. In a conventional workover, the Christmas tree is removed after the well is killed, and the preventers are bolted in place. There is a period during a conventional workover when the well is open to the atmosphere and controlled only by the kill fluid. This period occurs when the Christmas tree is being removed and the BOPs are installed.

After the preventers are in place and the well is killed, a small-diameter work string is run in the hole. A 1¼" EUE tubing string is normally used as a work string inside 2⅞" OD or 3½" OD tubing, and 1" EUE tubing with turned-down collars is usually used in 2⅜" tubing. Integral joint tubing is also used.

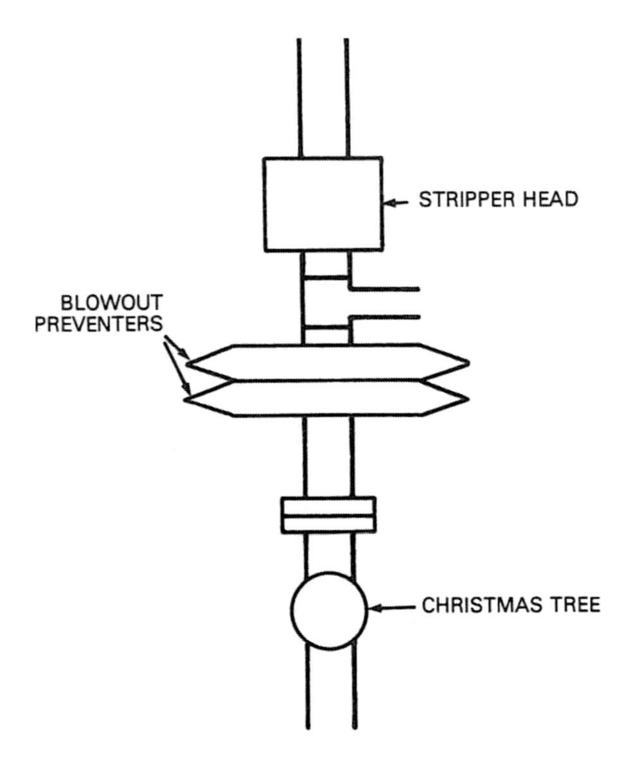

Figure 10.2. Blowout preventer installation for concentric-tubing workover

The equipment needed for a concentric-tubing workover, in addition to the tubing work string, is–

1. a lightweight hoisting and rotating unit;
2. blowout preventers;
3. a pump;
4. tanks for circulating fluid; and
5. small bits or mills if drilling or milling is required.

The equipment is essentially the same as for a conventional workover except that it is smaller and lighter – an advantage as well as a disadvantage. The lighter equipment is more adaptable to use on offshore platforms, and its use can lead to substantial workover savings.

But there are disadvantages. It is much more difficult to work inside 2⅜" OD or 2⅞" OD tubing than inside 7" casing. The smaller the area in which downhole work is performed, the greater is the potential for trouble. The space inside 7" casing is sufficient to perform all types of work safely. If a fishing job is necessary, it is possible to wash over the fish easily, since there is sufficient clearance. In 4½" OD casing, the equipment available does not have the same clearances, so fishing jobs are much more difficult. If the hole size is restricted to the ID of 2⅜" OD tubing, the problems are compounded. Nevertheless, concentric-tubing workovers using mini-conventional rigs or hydraulic snubbing units have many applications.

Coiled-Tubing Workovers

The coiled-tubing unit uses a continuous steel tube, normally ¾" to 1" OD, which is run into the well in one piece. Lengths of the tubing up to 16,000 feet long are stored on the surface on a reel in a manner similar to that of wireline. The unit is rigged up over the wellhead (fig. 10.3).

Friction blocks grip the tubing and push it into the well or pull it out. The tubing is injected through a control head that seals off the tubing and makes a pressure-tight connection. The coiled tubing can be injected against surface pressures up to about 5,000 psi, so it is not necessary to kill the well.

The coiled tubing can be run into the well at a fairly fast rate, 150 to 200 feet/minute. One

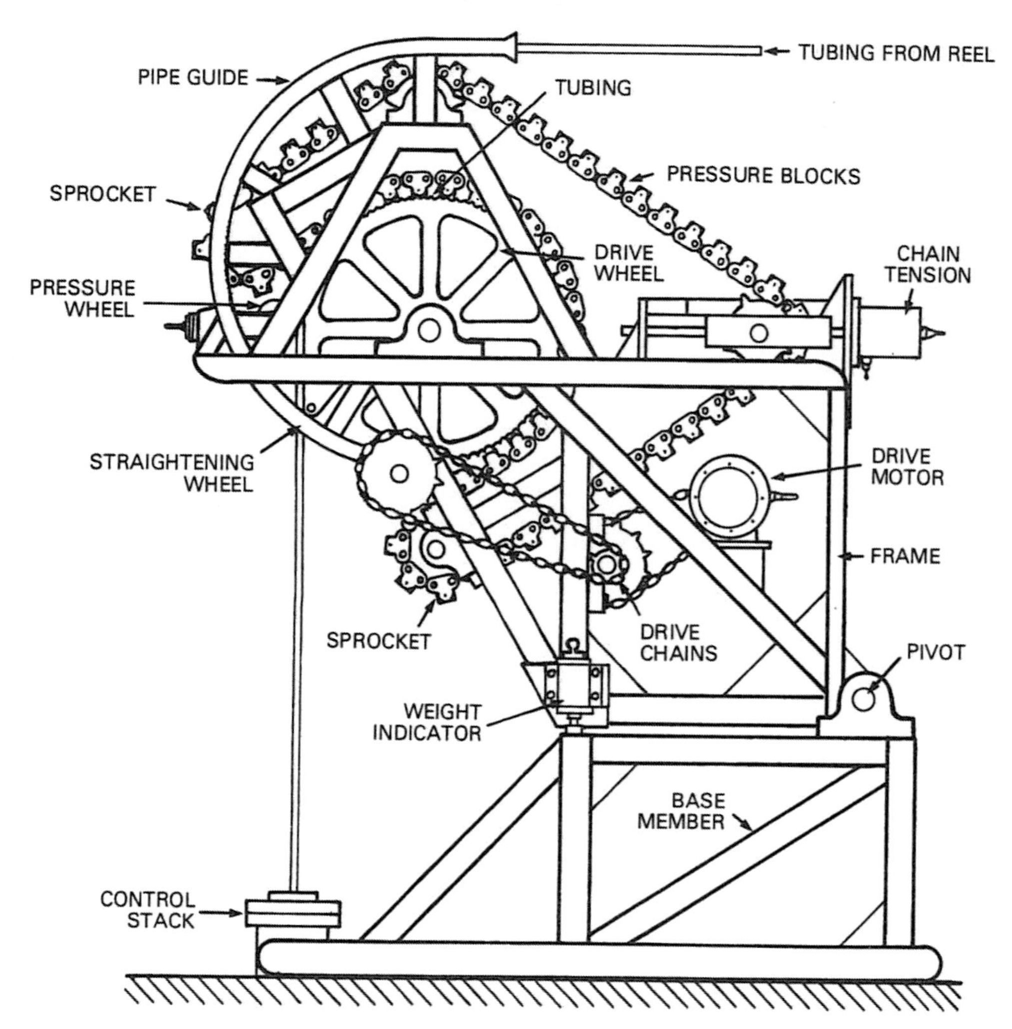

Figure 10.3. Typical coiled-tubing unit

unique feature of the coiled-tubing unit is that it is possible to circulate continuously while lowering the tubing in the hole. This feature is helpful when washing out sand fill. With concentric tubing, circulation is stopped each time a joint is added to the string.

Small-size hydraulic bits that permit cleanout operations to be carried out in a manner similar to a concentric-tubing operation are available.

The coiled-tubing unit is excellent for unloading wells, since gas or nitrogen can be injected continuously as the tubing is lowered in the well. The coiled-tubing unit is also useful in backwashing water-injection wells where the coiled tubing can be used to gas-lift the well and make it flow back to a pit.

The coiled-tubing unit has a number of disadvantages, however, that make it less desirable in many cases than the concentric-tubing method.

1. *Pressure drop:* One of the problems common to both coiled-tubing and concentric-tubing operations is the high pressure drop through the small-diameter tubing. The coiled unit has an added problem: it is necessary to pump through the tubing on the drum as well as through the tubing in the hole. If work is being done at 8,000 feet, it may be necessary to pump through 16,000 feet of coiled tubing.

2. *Coiled-tubing failure:* When coiled tubing is pulled from the well and wrapped around the drum at the surface, flex and stress occur in the

metal. After so many stress reversals, the tubing will fail. Fishing for a piece of parted coiled tubing can be very difficult. Good coiled-tubing operators keep track of the number of runs and replace the tubing periodically.

3. *Inability to spud:* It is not possible to spud coiled tubing to get past an obstruction as it is with concentric tubing.

Wireline Workovers

Wireline workovers are also performed through the tubing string. A lubricator is rigged up over the wellhead (fig. 10.4). The wireline tools can be inserted in the well through the lubricator under pressure. A variety of wireline tools is available. A bailer is available to bail out sand bridges. A dump bailer is available to set cement plugs in plug-back operations. Perforating guns are usually run on wirelines, and they can also be used.

Wireline work is usually performed with a line that varies in size from 0.072 inches to 0.1875 inches in diameter. The 0.1875-inch line is a braided electric line, while the lines from 0.072 to 0.105 inches are solid steel, or slick, lines.

A wireline reel and power unit is needed to store the line and provide the power to pull it from the hole. The wireline hoist unit has a measuring device and a weight indicator to permit measurement of depths. A lubricator is also needed to insert the wireline tools in the hole under pressure.

Wireline workovers are ideal for offshore operations, since they can be performed at a fraction of the cost of conventional workovers. The big drawback is that many types of workovers cannot be satisfactorily performed by wireline, and other methods must be used.

Through-Flowline Workovers

A system has been developed by Otis Engineering utilizing two tubing strings and pump-down plugs. It has its best application in highly deviated holes and ocean-floor completions.

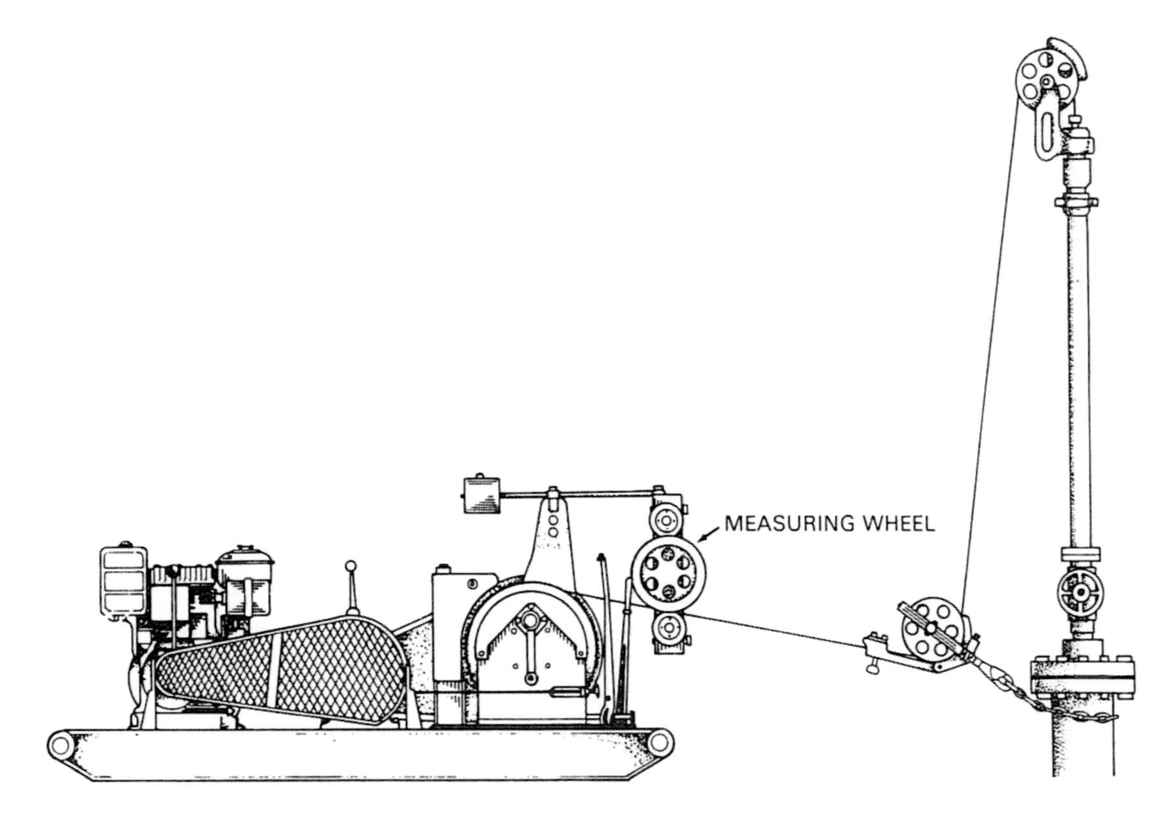

Figure 10.4. Typical wireline hookup (Courtesy of Halliburton)

Example Problems in Well Workover Methods

Sand Fill Problem

A typical problem that occurs in wells producing from unconsolidated sand formations is for the well to sand up (fig. 10.5). In the illustration, sand has filled up the hole to a point slightly above the perforations. The sand must be removed to leave the perforations unobstructed.

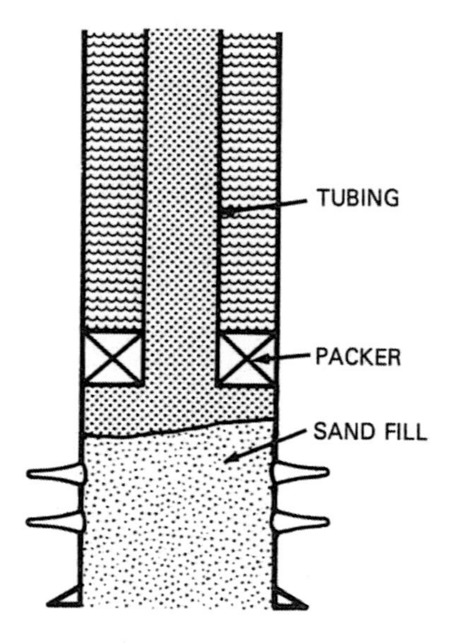

Figure 10.5. Sand fill problem

Conventional workover. A conventional workover to remove the sand fill requires that the packer and tubing be pulled from the hole. In order to accomplish this, the well must be killed by one of the three methods previously discussed. After the equipment is pulled from the hole, open-ended tubing is run in the well (fig. 10.6). The sand is removed by pumping water down the annulus as the tubing is slowly lowered into the sand. The water, together with the sand, is pumped up the tubing string.

A conventional workover of this type can usually be performed without difficulty. Since either 2⅜" OD or 2⅞" OD tubing is used, no difficulty is experienced in pumping at high-enough rates to remove the sand.

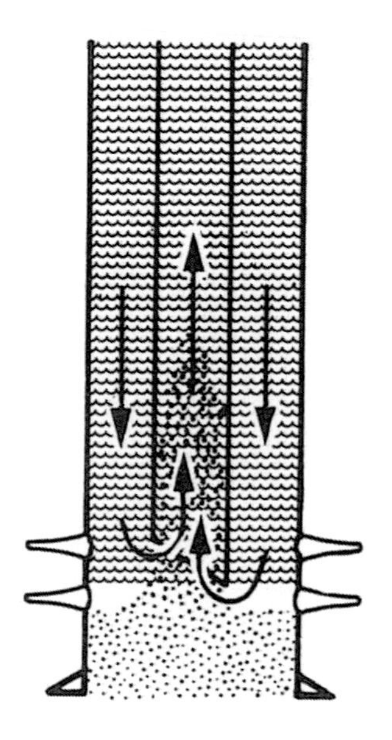

Figure 10.6. Washing sand in a conventional workover

After removing the sand fill, it may be desirable to run a gravel-packed screen and liner to prevent a recurrence of the sand problem. This can be very easily accomplished, and the largest size of screen and liner that will fit in the casing can be run. The use of the large screen and liner will result in maximum productivity for a screen-and-liner completion inside casing.

Concentric-tubing or coiled-tubing workover. The concentric-tubing and coiled-tubing workovers to remove sand are similar, so they will be discussed together. The well will probably be killed before either method is used. Although either method could be used under pressure, neither would be unless it were necessary.

When the concentric-tubing or coiled-tubing method is used, pulling the tubing and packer is not necessary. The coiled tubing or concentric tubing is run inside the production tubing string (fig. 10.7). Since the tubing is not pulled, a great deal of time is saved, especially with the coiled-tubing unit. The sand is circulated up the tubing by pumping down the tubing-tubing annulus.

The problem with the small-diameter tubing is that good circulation is hard to achieve. For example, pumping salt water down ¾" tubing at

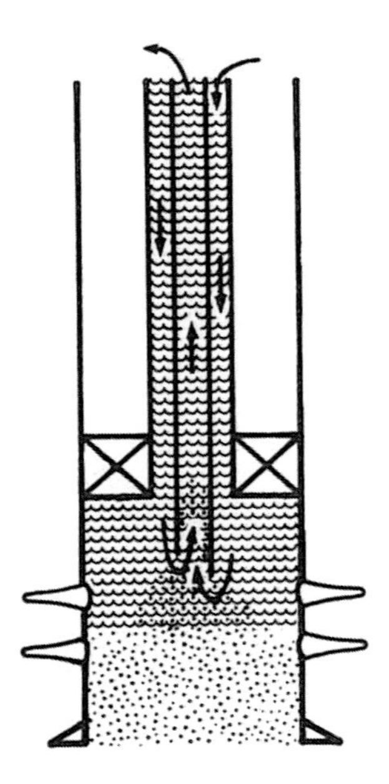

Figure 10.7. Sand removal with concentric tubing or coiled tubing

½ barrel/minute in a 9,000-foot well will require a surface pressure of 1,800 psi. Sometimes it is difficult to remove solids at the low circulation rates. To help in this situation, the water can be foamed by adding a foaming agent and nitrogen. Foams are capable of suspending pebble-size solids.

If it is desirable to run a screen and liner to minimize future sand problems, a concentric-tubing workover presents problems. The size of screen and liner that can be run is limited by the ID of the production tubing. In addition, it is not possible to run a perforation washer before the screen and liner are set. The result is decreased productivity as compared to that from the screen and liner set by a conventional workover.

Wireline workover. The simplest way to remove sand from a procedural standpoint is to use the wireline bailer. Unfortunately, it is usually not effective, especially if there is a very large sand fill. A wireline unit is rigged up on the well, and a sand bailer is run through the tubing string (fig. 10.8). Since a wireline unit is cheap and does not require pulling the tubing and packer, its use can sometimes be a cheap method for removing a small amount of sand fill.

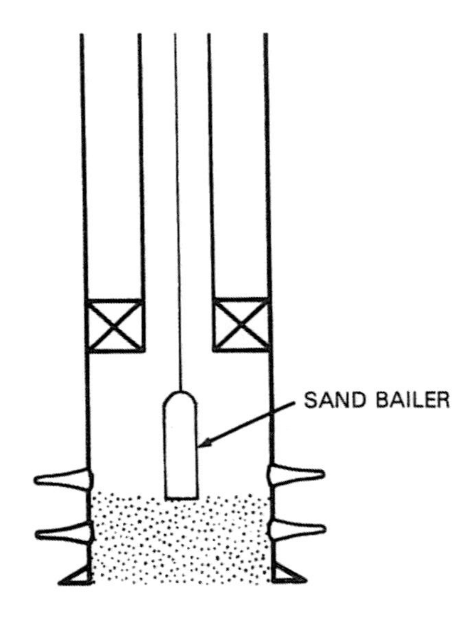

Figure 10.8. Sand removal by sand bailer on wireline

Squeezing Off Perforations and Recompleting

For the second example problem, a well has two potential producing intervals (fig. 10.9). The lower interval is depleted, so it is desired to squeeze off the lower perforations and perforate the upper zone.

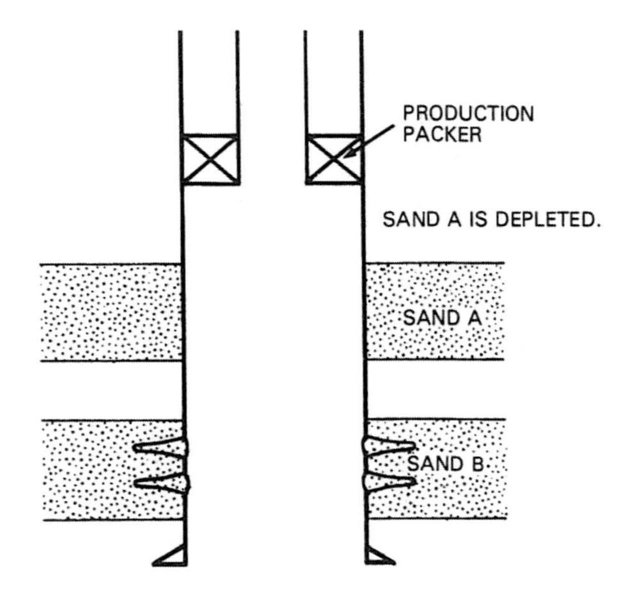

Figure 10.9. Well with depleted zone

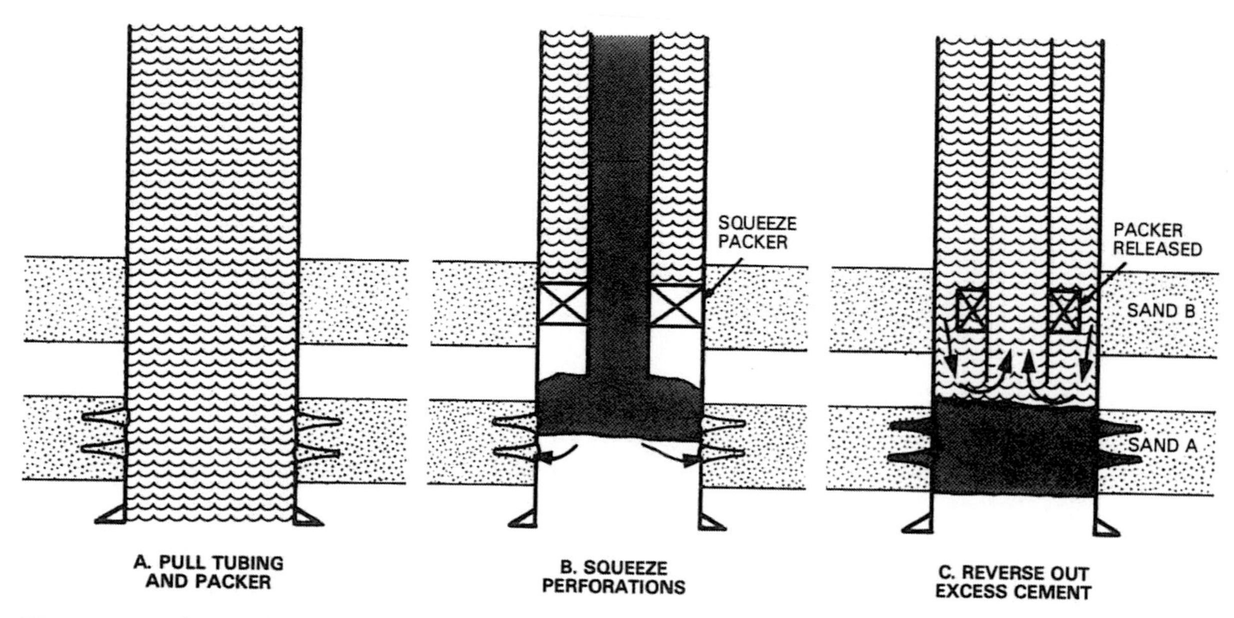

Figure 10.10. Conventional workover plug-back

Conventional workover. In a conventional workover (figs. 10.10 and 10.11), the first step is to kill the well and pull the tubing and packer. The lower perforations are then squeezed off (fig. 10.10*B* and 10.10*C*). It is not necessary to drill out the cement, since the lower zone is to be abandoned.

At this point the well needs to be perforated and the tubing and packer rerun. There are two options. The well can be perforated with a casing gun (fig. 10.11*A*), and after perforating, the tubing and packer rerun and the well placed on production. The tubing and packer can also be run before perforating, and the well perforated with a tubing gun (fig. 10.11*B*). The advantage of this method is that the well can be kept under control, yet still shot with a pressure differential to the wellbore.

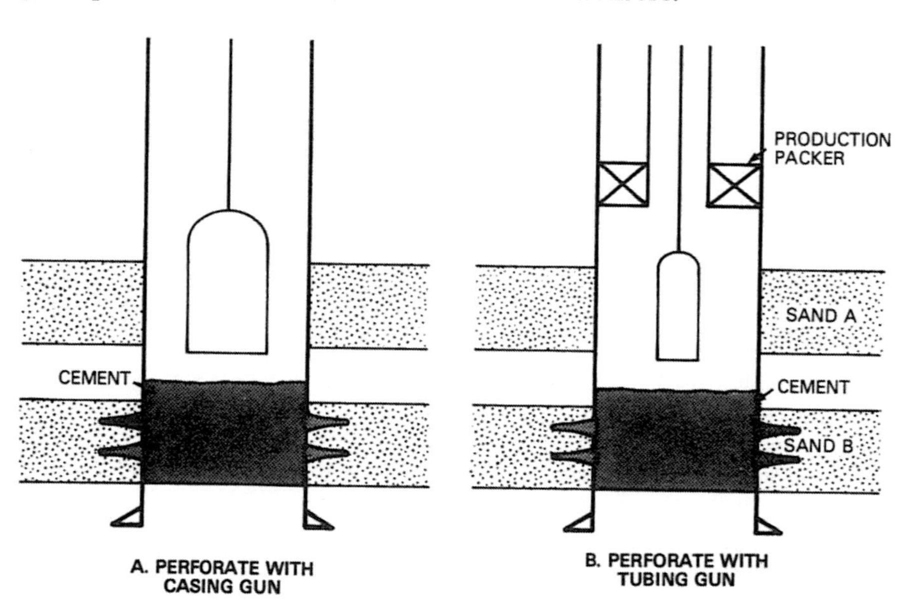

Figure 10.11. Conventional workover plug-back

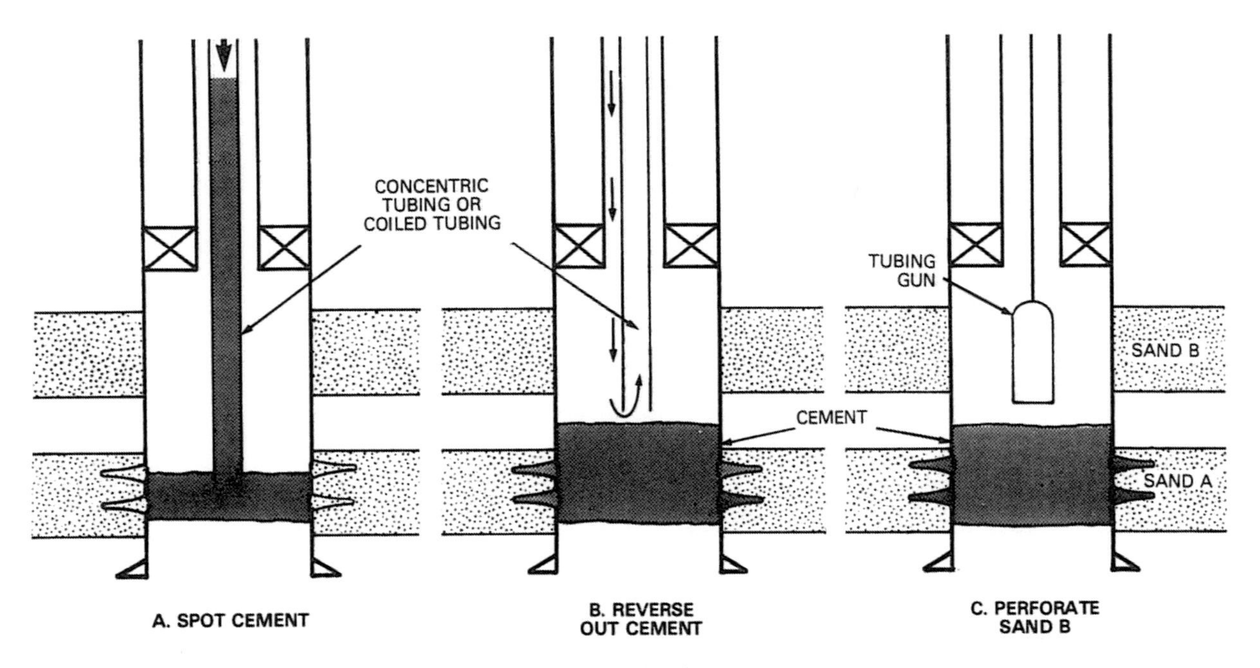

Figure 10.12. Concentric-tubing or coiled-tubing plug-back

Coiled-tubing or concentric-tubing plug-back. The plug-back can be performed with a coiled-tubing or concentric-tubing unit without pulling the tubing. The steps to be followed are shown in figure 10.12. Since it is not necessary to pull the production tubing, a saving in time and money can be realized, compared to a conventional workover.

Wireline workover plug-back. The wireline method is ideally suited to the plug-back. Since all of its operations are conducted through tubing, it is the cheapest method. The lower zone can be abandoned by placing a cement plug over the perforation zone by means of a bailer (fig. 10.13*A*). Two types of dump bailers are the electrically operated type and the frangible-disc

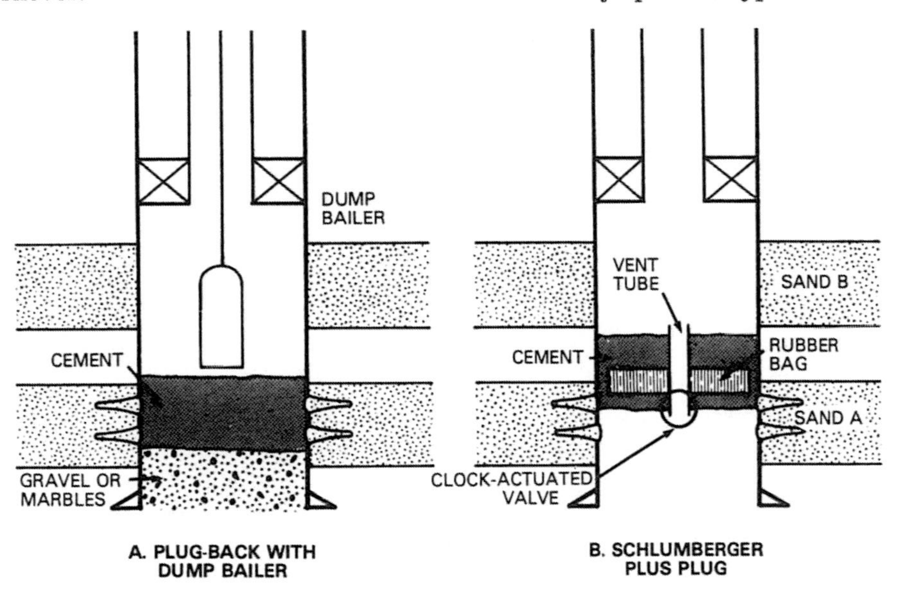

Figure 10.13. Wireline plug-back

type. The electrical dump bailer is opened and the cement dumped by firing a squib shot in the bottom of the bailer. It ruptures a soft metal disc and allows the cement to dump out of the bailer. The frangible disc is similar, except that the disc is ruptured by spudding down on a rod in the bottom of the bailer, allowing the cement to dump out of the bailer.

The bailer is run to the perforations and the cement dumped until there is sufficient fill to shut off the perforations. To minimize the amount of cement needed, marbles or coarse gravel can be dumped down the tubing string to partially fill up the hole.

One problem with the dump-bailer method is that many wells continue to build up bottomhole pressure for a long period after they are shut in. If they do, gas can percolate up through the cement and damage the cement plug before it fully sets up. In order to overcome this problem, the Schlumberger Plus Plug (fig. 10.13*B*) was developed. Since then, other companies have developed plugs with clock-operated bypass nipples.

The Plus Plug has a rubber bladder that is inflated with cement at the desired depth by firing the setting tool. The rubber bladder expands and makes a seal in the casing. In addition, some cement is left on top of the rubber bag. The important feature of the Plus Plug is that it has a clock-operated bypass nipple, which equalizes the pressure across the plug while it is setting. The clock closes the valve after a predetermined time based on the WOC time of the cement. Additional cement can then be dumped on top of the plug to improve its pressure-holding ability. After the well is plugged back by one of the two methods, the well is perforated through tubing (fig. 10.14).

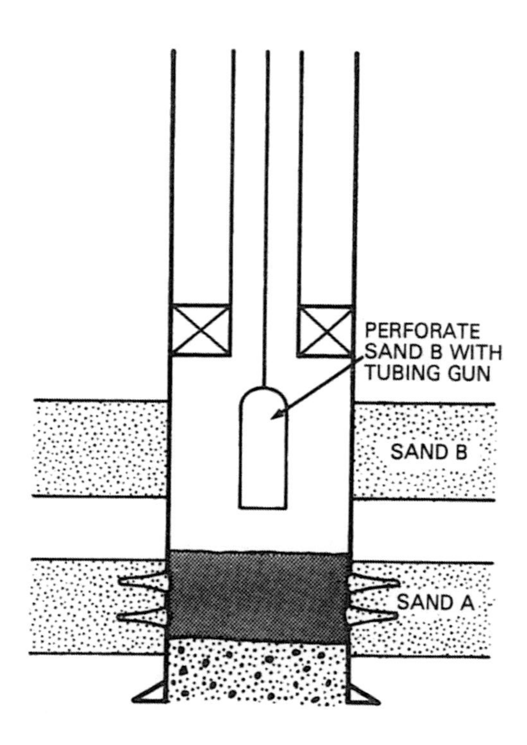

Figure 10.14. Perforating after wireline plug-back

Choosing the Optimum Workover Method

It is obvious from the brief discussion of workover methods that there is a considerable variation in cost with the different methods. Following is a list of the different methods in the order of ascending costs:

1. Wireline
2. Coiled-tubing
3. Concentric-tubing
4. Conventional

It is a good practice to consider all four methods in the order given whenever a workover is contemplated. If the wireline method is applicable, it should be used, since it is usually the cheapest. If a wireline workover is not feasible, then consideration should be given to coiled-tubing, concentric-tubing, and conventional workover, in that order. Each method has limitations, so the cheapest method that will get the job done should be selected.

Snubbing Units

Snubbing units are available to run or pull tubing under pressure. The hydraulic snubbing units are small and lightweight, making them easy to use offshore where deck space is at a premium and weight is important.

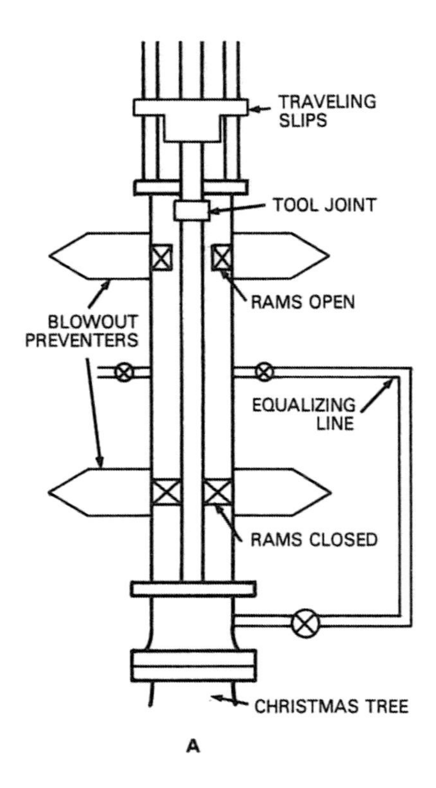

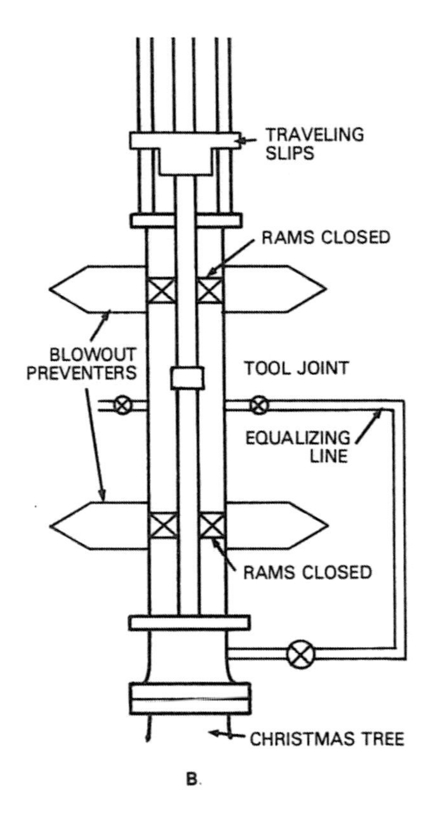

Principle of Operation

The operation of a hydraulic snubbing unit utilizing a concentric-tubing string is shown in a simplified drawing showing only two blowout preventers (fig. 10.15). High-pressure wells usually have two safety blowout preventers below the bottom preventer. For pressures below about 3,000 psi, stripper rubbers are used instead of the two blowout preventers.

In figure 10.15*A*, the hydraulic snubbing unit has been bolted to blowout preventers mounted on the Christmas tree. The snubbing unit has a set of traveling slips actuated by hydraulic cylinders that force the tubing into the hole or pull the tubing out of the hole. Note that the upper stripper rams are open, and the tool joint is being lowered below the upper stripper ram.

In figure 10.15*B*, the tool joint has passed through the upper stripper rams, and they have been closed. The equalizing valve is then opened to equalize the pressure between the two sets of stripper rams and the well.

In figure 10.15*C*, the lower stripper rams have been opened, and the tool joint has passed below them. The process is continued until the entire tubing string has been run into the hole.

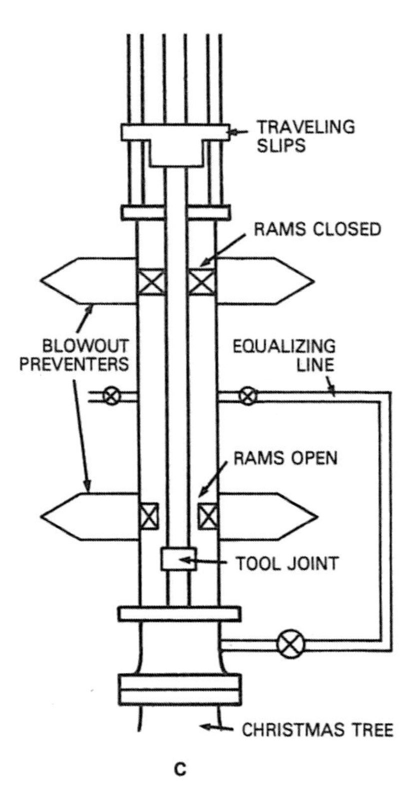

Figure 10.15. Hydraulic snubbing unit

Application

Hydraulic snubbing units can perform all of the tasks performed by mechanical rigs. A hydraulically powered rotary table is mounted on the traveling slip assembly to permit drilling. Although a hydraulic snubbing unit with concentric tubing is used to describe the principle of operation, the units can also be used to pull or run conventional tubing under pressure. The units can snub in pipe up to 5 ½" OD. A snubbing unit should be considered for both conventional and concentric-tubing workovers when it is not feasible to kill the well.

Workover Planning Considerations

After anomalous behavior has been determined for a well, the remedial measures usually involved in a workover are planned. Before considering individual problems that may exist, some broader aspects of workover planning should be examined.

Evaluation of Future Depletion of Wells

One of the points stressed in the well completion section of this text is that the initial completion should be adequate to deplete the well. Unfortunately, this is not always the case, and many wells are completed without provision for the recompletion to a different reservoir later in the life of the well. Also, conditions may have changed, and the original depletion plans are no longer valid. At any rate, workover planning should consider the future depletion of the well.

Depletion of single reservoirs. Some wells have only one productive reservoir, and no unusual problems should arise with this type of completion. If a workover is required, consideration should be given to how the well will be produced to depletion. Not many options are available, and everything should be fairly straightforward.

Depletion of wells with multiple reservoirs. The depletion of wells with multiple reservoirs is more complex and requires careful study to develop the optimum depletion plan. In some instances it may be possible to deplete all zones simultaneously by use of one of the multiple completion arrangements discussed in chapter 3. Sound reservoir management practices usually remove the option of commingling the zones. In addition, many regulating bodies do not permit commingled production of multiple reservoirs.

When a workover is performed, it is always important to determine whether the well should be converted to a multiple completion so that it can be depleted without additional workovers and recompletions.

Many wells are not candidates for multiple completions. The use of artificial lift by bottomhole electrical submersible pump is a good example of an instance requiring a single completion. It may be necessary to deplete each of several zones individually with single completions. In this event, it is desirable to deplete the lower zones first and then move up the hole to the upper zones. To understand why this is important, first note the two-reservoir situation in figure 11.1.

If sand *A* is depleted first, then it will have to be abandoned by squeezing off the perforations (fig. 11.2). After the perforations in sand *A* are squeezed off, sand *B* is perforated. Recall from the squeeze cementing discussion that it is difficult to get cement into all perforations. It is difficult and time-consuming to assure that all perforations are open so that they will receive cement during a squeeze job. Quite often it is necessary to resqueeze one or more times when the cement is drilled out so that the well can be produced below squeezed-off perforations.

Another problem with squeezed-off perforations above the producing interval is the

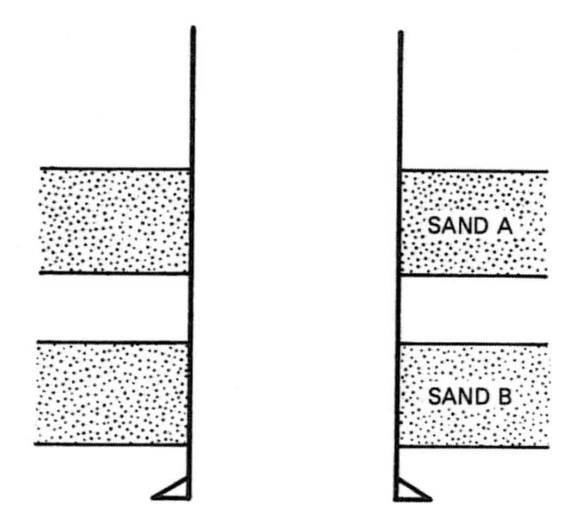

Figure 11.1. Two-reservoir completion

possibility that a successful squeeze job may break down in the future. This can be particularly hazardous if the squeezed-off zone is a water drive gas reservoir. The author once observed a serious blowout that occurred when squeezed-off upper perforations in a depleted water drive gas sand broke down. The well flowed gas, water, and sand for several days before it bridged over and well control was established. A depleted water drive gas sand still has a high pressure, and it can flow water and gas even though commercial gas production is not possible.

Even if the breakdown of the perforations does not present a potential blowout hazard, it is still undesirable. It may result in extraneous water production, or the abandoned zone may act as a "thief" zone and take oil or gas from the producing zone.

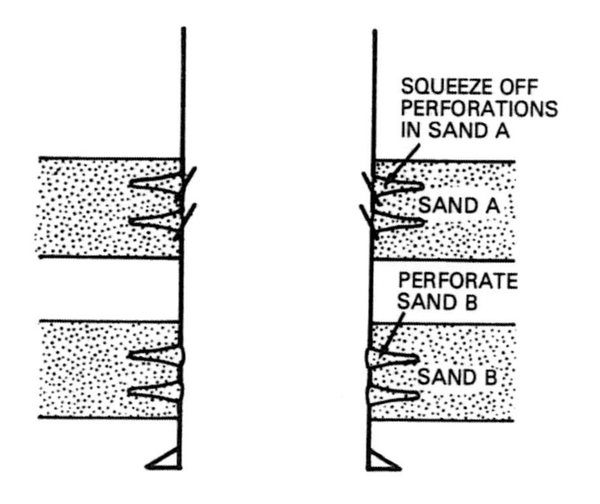

Figure 11.2. Abandoning upper zone in two-zone completion

If sand *B* is squeezed off first and the well recompleted to sand *A* (fig. 11.3), no problem should occur even if some of the perforations are plugged and do not accept cement.

Note that in the situation in which the lower perforations are squeezed off first, it is not necessary to drill out or reverse out the cement opposite the lower perforation. Since solid cement is in the casing, a seal of the bottom zone will normally be obtained even if cement does not go into all of the perforations.

It is obvious why it is advisable to start from the bottom zone and work up the hole to deplete a well, but this course is not always feasible. One of the reasons may be competitive position.

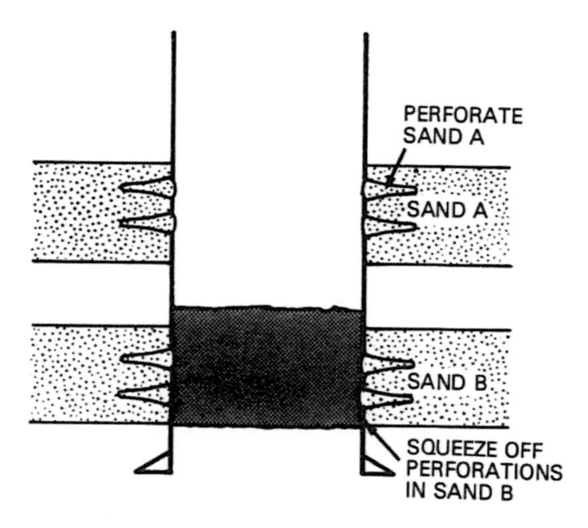

Figure 11.3. Abandoning lower zone in two-zone completion

Evaluation of Competitive Position

One of the reasons a well cannot be depleted from the bottom up may be the problem of competitive position. In some countries where an entire reservoir is operated by the same operator, the problem does not exist. Many areas, however, do have competitive reservoirs and should be given attention by the workover planner.

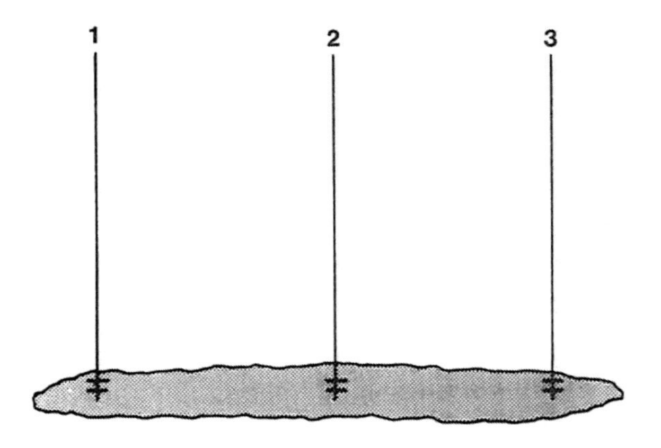

Figure 11.4. Dissolved-gas drive oil reservoir

Dissolved-gas drive oil reservoirs. First consider a dissolved-gas drive oil reservoir similar to the one shown in figure 11.4. Wells in such a reservoir will have ultimate oil recoveries approximately proportional to their current producing rates. The concept is called a "straw-in-the-pot," and its use gives reasonable results. If two people are sucking on straws from a common soda water bottle, they will get an amount of soda proportional to how hard they suck. The same principle holds for oil reservoirs. As an example, make the following assumptions for the reservoir (fig. 11.4).

Total remaining reserves in field = 105,000 barrels.

Well No.	*Current Producing Rate, BOPD*
1	300
2	700
3	500
	1,500

Well No.	*Remaining Reserves Expected, Barrels*
1	21,000 (300/1,500 × 105,000 = 21,000)
2	49,000 (700/1,500 × 105,000 = 49,000)
3	35,000 (500/1,500 × 105,000 = 35,000)
	105,000

Each well will share in future reserves in accordance with its percentage of the total field production. This proportion is an approximation, of course, but it is a good one and is used frequently in reservoir engineering to allocate field reserves to individual wells. It assumes that none of the wells will be stimulated without stimulating them all, a very reasonable assumption.

Water drive oil reservoirs. Next consider an edgewater drive, high-relief oil reservoir (fig. 11.5). The "straw-in-the-pot" concept still holds, but consideration must be given to how the reservoir structural position affects each well's recovery. In the edgewater drive reservoir, pistonlike displacement of the oil by water can be assumed. In other words, when the water level reaches well no. 1, it will cease to produce oil. This assumption is reasonable, since edgewater drive wells normally water out very quickly and produce water for a relatively short time. To simplify the approach, assume that all three wells are producing at the same rate.

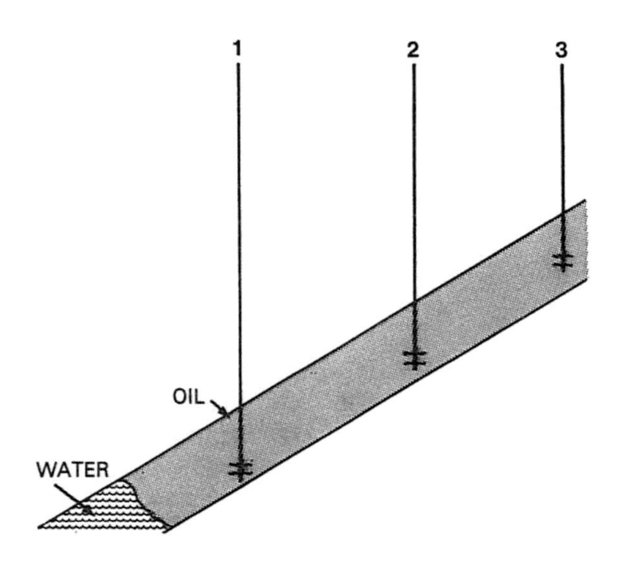

Figure 11.5. High-relief edgewater drive oil reservoir

It can readily be seen that all three wells will share equally in the oil produced while water moves through section *A*. Therefore, each well will recover an amount equal to *A*/3 while section *A* is depleted. As the water moves through section *B*, wells 2 and 3 will share equally in the production, since well no. 1 has watered out. Therefore, no. 2 and no. 3 will each recover *B*/2.

When the water level gets to well no. 2, only well no. 3 will still be producing, so it will get all of the oil from section *C*. Summarizing,

Well No.	*Recovery*
1	$A/3$
2	$A/3 + B/2$
3	$A/3 + B/2 + C$

The example was simplified by assuming that the three wells had equal producing rates. What if the wells produced as follows?

Well No.	*Producing Rate*
1	150
2	450
3	300
Total	900

In this case the per-well recovery will be–

Well No.	*Recovery*
1	$A \times 150/900 = A/6$
2	$A \times 450/900 + B \times 450/750 = A/2 + 3B/5$
3	$A \times 300/900 + B \times 300/750 + C$ $= A/3 + 2B/5 + C$

Competitive position in interval selection. Returning to consideration of the order in which a well with multiple reservoirs is depleted, consider the case shown in figure 11.6. Assume that sand *A* is a competitive, water drive reservoir and requires a bottomhole submersible pump to deplete. Since a submersible pump requires a single completion, sand *A* must be depleted as a single completion. Further assume that sand *B* and sand *C* are noncompetitive–that is, no other well is producing from these sands. Depleting sand *C* and sand *B* before sand *A* will result in lost reserves from sand *A*. While sands *B* and *C* are being depleted, other operators will have their "straws-in-the-sand *A*-pot," and they might even deplete it before recompletion to sand *A*. In a situation like this, sand *A* should be depleted first to ensure getting a fair share, and then sands *B* and *C* can be depleted.

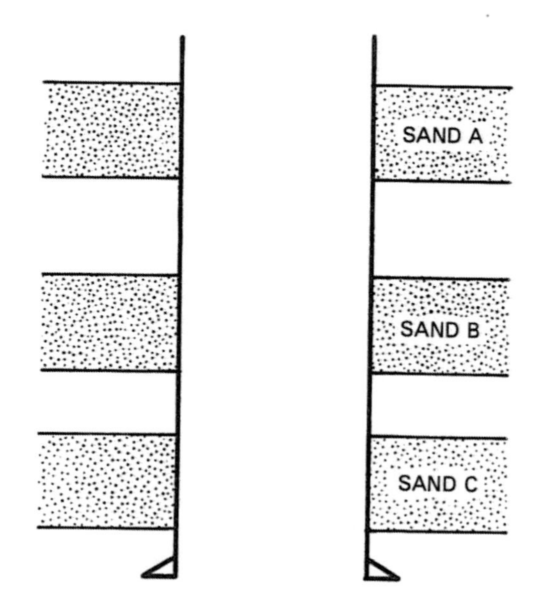

Figure 11.6. Depletion of multiple-reservoir competitive sands

Workovers to Reduce Water Production

In many instances wells in bottom-water drive reservoirs produce water prematurely because they are completed too close to the water-oil contact (fig. 11.7). Often the water production can be shut off by squeezing off the perforations

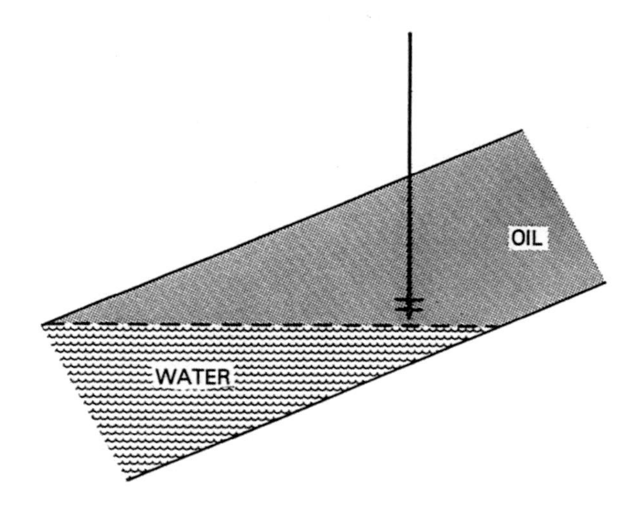

Figure 11.7. Water drive well completed near water-oil contact

and then perforating higher in the section. Eventually the well will make water as the water level rises with continued production.

Sometimes the well makes water prematurely because it is produced at too high a rate and the water cones up from below (fig. 11.8). Usually very little can be done to remove a water cone once it has developed. The best procedure is to squeeze off the current perforations and reperforate as high in the oil zone as possible. The production rate should also be reduced to minimize future coning.

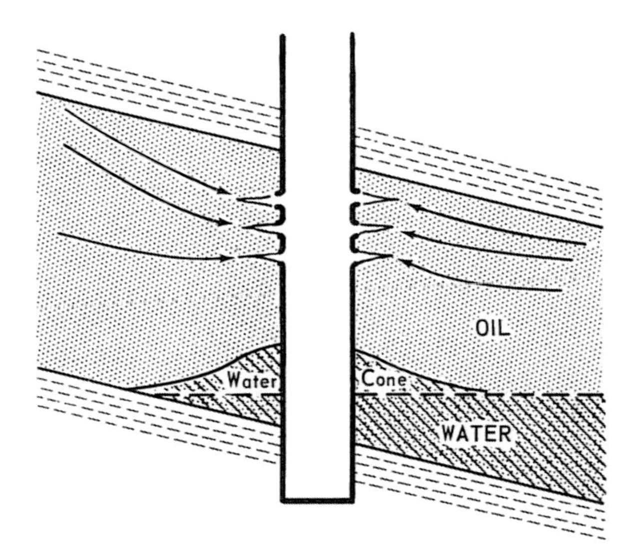

Figure 11.8. Water coning (Copyright 1972 by SPE-AIME)

Workovers to Reduce Gas Production

If a well is completed too close to an expanding gas cap, it may produce gas prematurely (fig. 11.9). The gas can usually be shut off by squeezing off the existing perforations and perforating lower in the well. Since the GOC will move down with production, the well will eventually start making gas again.

Sometimes the gas production is caused by producing the well at too high a rate and the gas cones down (fig. 11.10). Gas cones are very difficult to remove. If the well is shut in, oil will normally not move back into a gas-invaded section. Sometimes an *oil squeeze* will help. In an oil

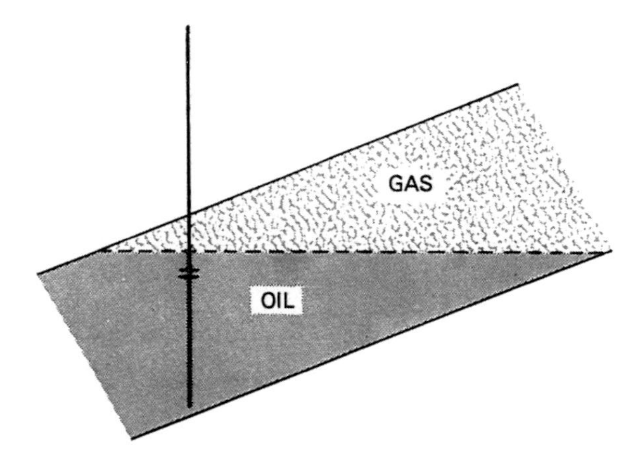

Figure 11.9. Well completed near gas-oil contact

squeeze, oil with a surface tension reducing agent is pumped into the perforations in an attempt to resaturate the gas-invaded area with oil. The permeability to gas will be reduced. Unfortunately, as in the case of a water cone, very little can be done to correct gas cones. If the wells are produced at the same rate, gas coning will recur. The best solution is to shut in wells near the gas cap and produce oil from wells further down. The reverse would be true for a water drive field with water-coning problems.

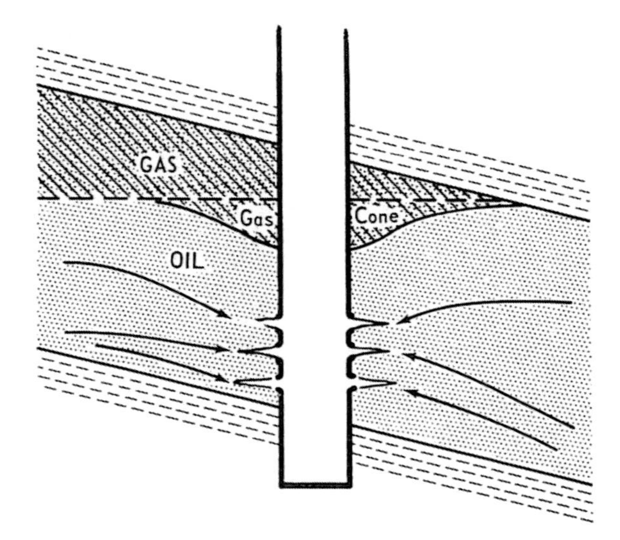

Figure 11.10. Gas coning (Copyright 1972 by SPE-AIME)

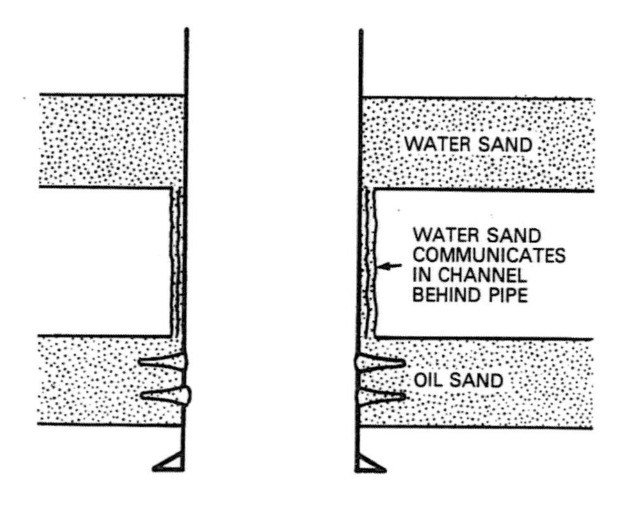

Figure 11.11. Extraneous water production

Workovers to Eliminate Extraneous Water Production

Some wells produce water that migrates behind the pipe from an extraneous zone because of a poor cement job. A sudden large increase in water production from a dissolved-gas drive oilwell usually indicates that the water is from another zone (fig. 11.11). The problem can be solved by using the circulation cement-squeeze technique. In this technique the water sand is perforated, and the tubing on a packer is set between the two zones (fig. 7.12). Cement is circulated through the channel. In this type of repair, if circulation can be established between the two sets of perforations, then the job can usually be successfully performed. The failures are often due to inability to circulate once the water sand is perforated. A cement bond log is helpful in determining whether circulation can be established and where to perforate.

Special Problems with Workovers

Some special situations have to be considered in planning a workover. If the bottomhole pressure is very low, restoring a well to production after it is killed with fluid may be difficult, especially true for low-pressure gas wells. In these wells it is better to select a method that will not require killing the well. A wireline workover, coiled-tubing techniques, or a snubbing unit to pull the tubing under pressure should be considered.

Some wells are damaged by fluid regardless of the type of fluid selected. Here again, the wireline, coiled-tubing, or snubbing unit workover techniques should be considered.

Selection of Workover Method

The various workover methods available were discussed in chapter 10. The least expensive method to do the job is usually the one to select. Use of the various workover techniques should be considered in the following order:

1. Wireline
2. Concentric-tubing or coiled-tubing
3. Conventional

The use of a snubbing unit should be considered with either conventional or concentric-tubing workovers.

Writing Workover Procedures

Once a well problem has been isolated by problem well analysis, and the type of remedial measure selected, the next step is to write a workover procedure. A workover should never be started without a written step-by-step procedure outlining exactly what is to be performed. A good procedure should contain the following data.

Basic Well Data

Make a list in tabular form of all pertinent equipment installed in the well. Give casing size, grade, weight, setting depth; tubing size, grade, weight, setting depth; type and location of packers; wellhead type, size, and arrangement. The reference point or datum for wellbore measurements should be included. Furnish enough data so that the job can be safely performed without going to the well file for additional information.

Wellbore Sketch

The wellbore sketch should include some of the basic well data that is also included in tabular form under "Basic Well Data." The wellbore sketch gives a graphical representation of the wellbore configuration, showing previous producing intervals, location of packers, tops of producing zones (fig. 9.9). Wellbore sketches are valuable in the planning of a workover and in the execution of the workover plans. They show at a glance what is in the well, potential pitfalls, and other essential information.

Pertinent Well History

Give a brief history of previous work performed, such as squeezed-off perfs above the current producing interval or any other work that will affect current operations.

Reason for Workover

Reason for the workover may be obvious to the person preparing the brief, but the field man in charge of performing the workover needs to know why a particular operation is being performed in order to carry out the procedure effectively. In addition, explaining why the workover is being performed permits a better post-workover evaluation. This reason is especially important if another person later evaluates the workover. Include a concise but adequate reason. For example, "to increase production" is not enough.

Current Well Tests

Current well test data are needed to compare with post-workover tests to evaluate the workover. They are also valuable for workover studies made at a later date.

Procedure to Be Followed

A step-by-step procedure is needed to assure that the workover is performed as desired. For maximum effectiveness, prepare the procedure to show important steps and leave details to the field man.

Well-Testing Procedure after Workover

List the tests that are desired to evaluate the workover properly. If any special testing such as pressure buildups are desired, include detailed procedures for them in the workover procedure.

XII Beam Pumping In Artificial Lift

Artificial lift is required to deplete most oilwells. Over 90% of the wells in the United States are currently being artificially lifted. Beam pumping is the most commonly used method, accounting for over 85% of all artificial lift installations, and will be the focus of discussion here. Gas lift, bottomhole centrifugal pumps, and bottomhole hydraulic pumps, in this order, are the other artificial lift methods used.

The Beam Pumping System

The beam pumping system is mechanically very simple, consisting of a surface unit that transmits up-and-down motion to a bottomhole pump by means of a sucker rod string (fig. 12.1). Although the system is simple, many factors are involved in its proper design. Formulas have been developed over the years to be used in designing an optimum pumping system. They will be discussed and illustrated by example problems. Unfortunately these formulas are not exact, and experience is still a key factor in good design.

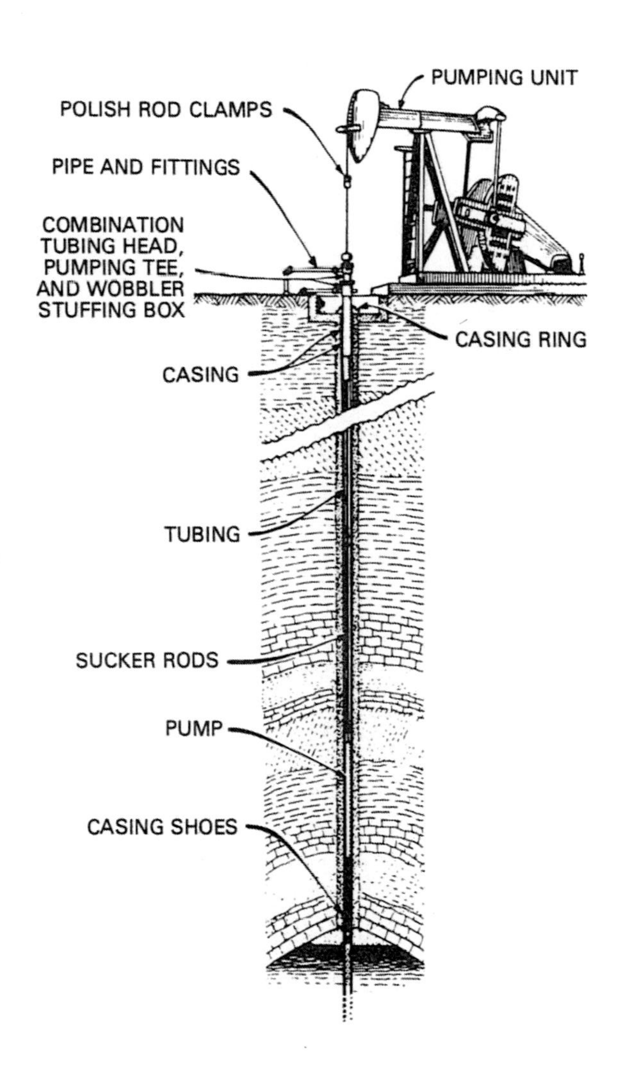

Figure 12.1. Beam pumping system. From Kermit E. Brown, *The Technology of Artificial Lift Methods,* vol. 2a (Tulsa; PennWell Publishing Company, 1980).

Surface Equipment

The most commonly used beam pumping unit (fig. 12.2) is a conventional, or class I lever, system, the first type of unit to be developed.

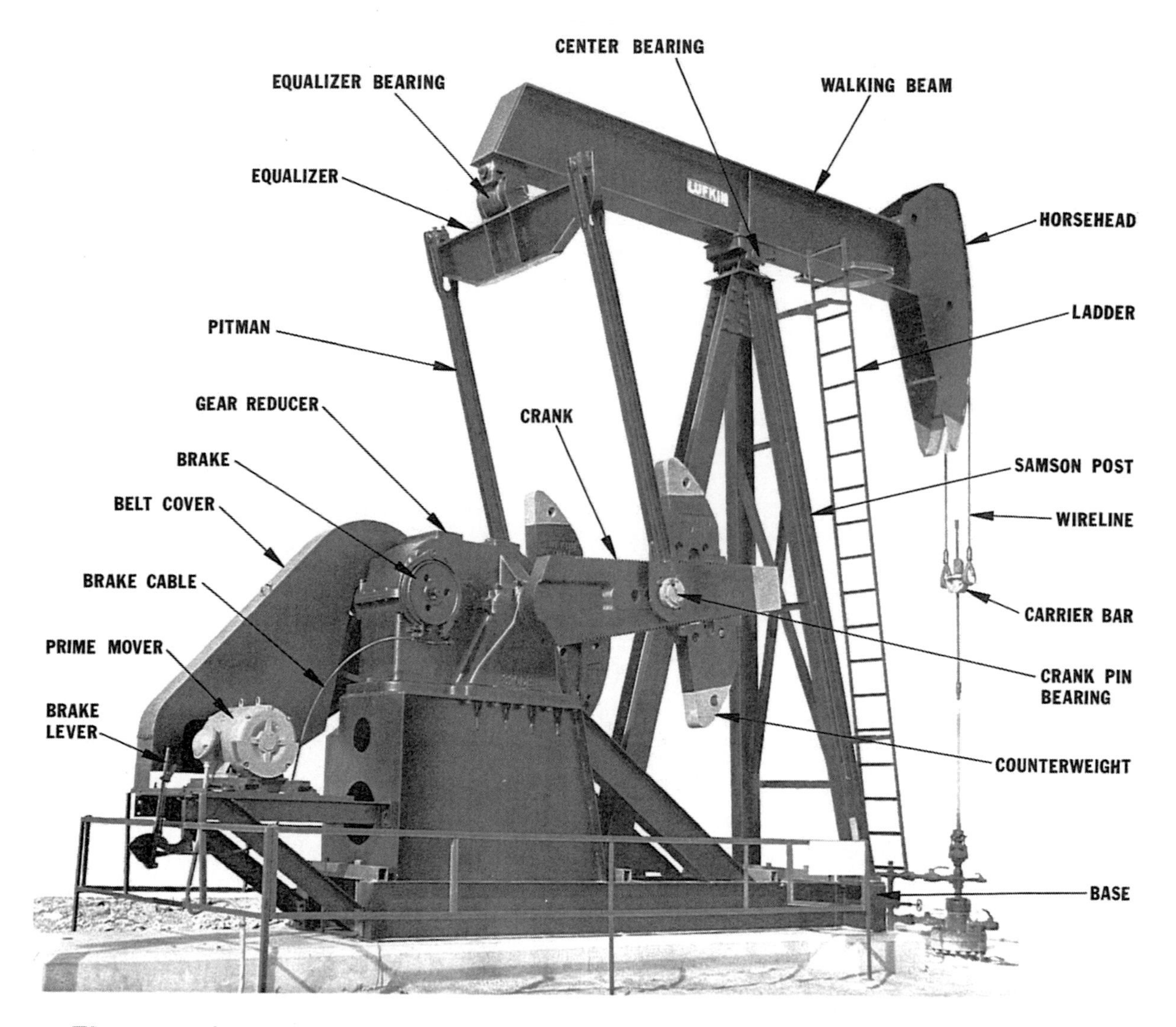

Figure 12.2. Conventional pumping unit, class I lever system (Courtesy of Lufkin Industries)

The fulcrum for the walking beam is located between the horsehead and the pitman connection. A prime mover, either a gas engine or an electric motor, drives the unit through a gear reduction box. A crank arm converts the rotary motion of the gear box to a reciprocating up-and-down motion, which is transmitted to the walking beam by means of the pitman arm. The up-and-down motion is transmitted to the rod string by the movement of the walking beam. The rod string is attached to the walking beam by a bridle attached to the horsehead. The horsehead and bridle arrangement provide for a vertical, not angled, pull on the rod string.

Counterweights are attached to the crank arm to balance the weight of the rods. The counterweights minimize the amount of horsepower required for pumping by balancing the weight of the rods and part of the fluid weight.

Another popular type of beam pumping unit is the class III lever system (Lufkin Mark II), in which the pitman is located between the fulcrum of the walking beam and the horsehead (fig. 12.3).

The factors that must be considered in the selection of the beam pumping unit and associated surface equipment are as follows:

Structural rating of unit: The unit must have sufficient capacity to carry the weight of the sucker rods and the fluid being lifted. The dynamic loads of the rods and fluid must be considered.

Torque capacity of gear box: The gear box must have sufficient torque capacity to transmit the power needed to lift the fluid at the desired rate.

Horsepower rating of prime mover: A prime mover of suitable size must be selected.

Downhole Equipment

The reciprocating motion of the pumping unit is transmitted to the subsurface pump by means of sucker rods. The sucker rods must be of the proper diameter and made of a material with sufficient strength to transmit the loads imposed by the pumping action. The pump must be of the proper size and design to pump the amount of fluid desired.

Figure 12.3. Class III lever system (Courtesy of Lufkin Industries)

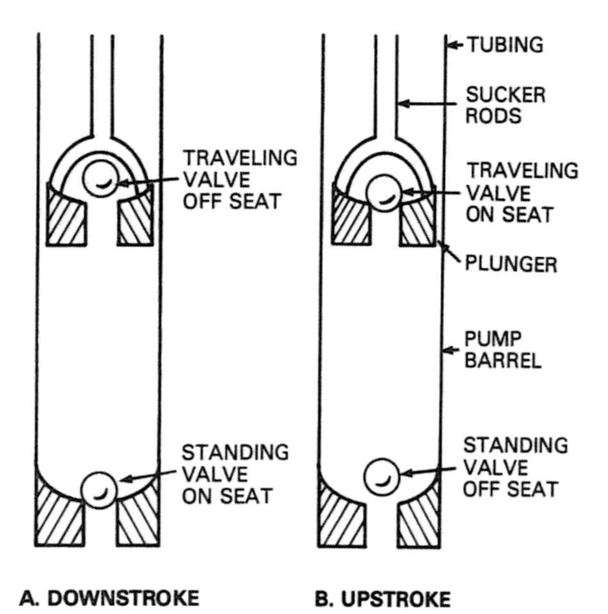

Figure 12.4. Subsurface pump operation

Subsurface Pumps

The schematic drawing of a subsurface pump (fig. 12.4) shows a typical positive-displacement, reciprocating pump. The pump's essential parts are plunger, barrel, traveling valve, and standing valve. The standing and traveling valves are usually spherical steel balls with precisely ground seats.

Operation

On the downstroke (fig. 12.4*A*), the standing valve is closed because of the weight of the fluid in the tubing. The traveling valve is open. At the bottom of the stroke, the weight of the fluid in the tubing causes the traveling valve to close. As the pump starts its upward motion (fig. 12.4*B*), the pressure in the barrel below the plunger is reduced, opening the standing valve and allowing fluid to flow into the barrel. As the plunger moves upward, the fluid in the pump above the plunger is pumped to the surface. After the pump reaches the top of the stroke and starts down, the fluid load is shifted to the standing valve, and it closes. The traveling valve opens as it moves downward in the fluid that is being supported by the standing valve.

Pump Displacement Calculations

It will be demonstrated later that for a given depth and volume of fluid there is an optimum pump size. For the present, the calculations necessary to determine pump displacement will be examined. Since oilfield production volumes are measured in barrels per day, and pump plunger diameters and plunger strokes are measured in inches, the common practice is to use a pump constant, or K factor, to determine pump displacement.

The theoretical displacement, PD, of a pump is the area of the plunger, Ap, in square inches, times the plunger stroke length, Sp, in inches.

$$PD = Ap \times Sp$$

This equation gives the pump displacement in cubic inches/stroke. Since an answer in barrels per day is desired,

$$PD = \frac{Ap \times Sp \times N \times 1{,}440}{9{,}702}$$

where

1,440 = minutes/day
9,702 = cubic inches/barrel
N = strokes/minute.

Since the pump constant $K = 0.1484\ Ap$,

$$PD = K\,Sp\,N.$$

Tables of K values for different pump sizes are available to facilitate displacement calculations. Table 12.1 lists K values for commonly used pump sizes.

TABLE 12.1
PUMP DATA

Plunger Diameter (inches)	Area of Plunger, A_p (square inches)	Constant, K
$1\frac{1}{16}$	0.866	0.132
$1\frac{1}{4}$	1.227	0.182
$1\frac{1}{2}$	1.767	0.262
$1\frac{3}{4}$	2.405	0.357
2	3.142	0.466
$2\frac{1}{4}$	3.976	0.590
$2\frac{1}{2}$	4.909	0.728
$2\frac{3}{4}$	5.940	0.881
$3\frac{3}{4}$	11.045	1.640

Rod and Tubing Effects

It may seem that designing a correct pumping unit installation is simple, since all that is necessary is to determine how much fluid is desired and then pick a pump diameter, stroke length, and number of strokes per minute to produce at the desired rate. Unfortunately, rod and tubing stretch occurs because of the fluid loads imposed. Since the weight of the fluid load varies with the plunger diameter, increasing the plunger diameter also increases the rod and tubing stretch. These two changes tend to offset each other. If the plunger diameter is increased to get more pump displacement, the increased fluid load will cause more rod stretch, which will decrease effective pump displacement.

A factor that helps to increase the effective stroke length at the pump is plunger overtravel. Plunger overtravel is caused by the elongation and contraction of the rod string due to the dynamic loads imposed by the pumping cycle.

Tubing Stretch

In order to understand tubing stretch, it is necessary to consider the action of the bottomhole pump again. Refer again to figure 12.4. It is obvious that on the upstroke of the pump, the fluid load in the tubing is supported by the rod string, since the traveling valve is closed. At the beginning of the downstroke, the fluid load is transferred from the rod string to the tubing string as the standing valve closes and the traveling valve opens. The extra weight imposed by the fluid load causes the tubing to stretch. The stretch of the tubing effectively shortens the plunger stroke. This process can be visualized by considering that on the upstroke the tubing tends to follow the plunger up the hole as the fluid load is picked up by the rod string, reducing the effective plunger stroke.

The amount of tubing stretch can be calculated by the use of Hooke's law. Hooke's law can be repeated as –

$$\Delta L = \frac{FL}{EA}$$

where

ΔL = change in length of object
F = force imposed on object
L = length of object
E = Young's modulus = 30×10^6 for steel
A = cross-sectional area of object.

If E_t = tubing stretch in inches and
W_f = weight of the fluid in pounds, then –

$$E_t = \frac{12\ W_f \times L}{E \times A_t}$$

where

L = length of tubing in feet
A_t = cross sectional area of the tubing, square inches.

The amount that tubing will stretch can be illustrated by the following example:

Assume that salt water with a specific gravity of 1.08 is being pumped with a 2¼" plunger set in 2⅞" OD, 6.5 lb/ft tubing at a depth of 5,000 feet.

$$W_f = 5{,}000 \times 1.08 \times 0.433 \text{ psi/ft} \times A_p$$

where

$A_p = 2.25^2 \times \pi/4$
W_f = 9,297 lb

$$E_t = \frac{12 \times 9{,}297 \times 5{,}000}{30 \times 10^6 \times 1.812} = 10.3''.$$

If the well were being pumped with a 54" surface stroke, 19% of the stroke would be lost due to tubing stretch.

Tubing stretch can be prevented by using a tubing anchor. This is a device similar to the slip section of a tension packer. Tubing anchors also help prevent tubing or casing wear. The problem with tubing anchors is that they may be hard to retrieve in corrosive wells or wells that make sand.

Rod Stretch

The rod string experiences a problem similar to that of the tubing because of the fluid load. Refer again to figure 12.4; it can be seen that the rod string picks up the fluid load at the start of the upstroke. When the rods start up, the plunger doesn't move until the stretch is taken out of the rod string, shortening the effective plunger stroke.

If E_r = rod stretch, inches
A_r = rod cross-sectional area, square inches

$$\text{Then } E_r = \frac{12W_fL}{EA_r}$$

For tapered rod strings,

$$E_r = \frac{12 \times W_f}{E}\left(\frac{L_1}{A_{r1}} + \frac{L_2}{A_{r2}} + \frac{L_n}{A_{rn}}\right)$$

where

L_1 = length of rod A_{r1}
L_2 = length of rod A_{r2}.

By referring again to the same example problem used to illustrate tubing stretch, it can be demonstrated how much rod stretch occurs.

Assuming 7/8" sucker rods (cross-sectional area = 0.601 in.^{2}),

$$E_r = \frac{W_fL}{EA_r} = \frac{12\times 9{,}297 \times 5{,}000}{30 \times 10^6 \times 0.601} = 30.9\text{"}.$$

The combined rod and tubing stretch for the example problem is

$$E_{total} = E_r + E_t = 10.3 + 30.9 = 41.2\text{"}.$$

If a surface stroke of 54" were used, then the rod and tubing stretch would reduce the effective stroke by 76%. Obviously a 54" surface stroke would not be practical with the pump and rod combination used in the example.

Plunger Overtravel

As mentioned previously, plunger overtravel increases the effective plunger stroke. Plunger overtravel is caused by the elongation of the rods due to the dynamic forces imposed during the pumping cycle.

Plunger overtravel can be illustrated by referring to the example used to demonstrate rod and tubing stretch. In the example there are 5,000 feet of 7/8" sucker rods. Since 7/8" sucker rods weigh 2.25 lb/ft, the total weight of the rods, W_r, is

$$W_r = 5{,}000 \times 2.25 = 11{,}250 \text{ lb.}$$

If the well is pumped at a speed (*N*) of 20 strokes per minute, then the 11,250 lb of rods will be traveling at a speed of 7.1 feet/second as the plunger reaches the bottom of the stroke. In about three-fourths of a second the 11,250 lb of rods are brought to a halt and start back up the hole. Obviously the 11,250 lb of rods want to continue past the bottom of the stroke, and they do. The downward dynamic force stretches the rod string and causes a longer plunger travel. This phenomenon is called *overtravel.* The same thing happens at the top of the stroke. In 1931, Marsh and Coberly proposed the following formula to account for plunger overtravel:

$$E_{ot} = (1.93 \times 10^{-5})\, S \left(\frac{L}{1{,}000}\right)^2 N^2$$

Since the formula did not account for tapered rod strings, utilizing two or more sizes of rods, the following revision of the Marsh-Coberly formula became generally used:

$$E_{ot} = 1.55\left(\frac{L}{1{,}000}\right)^2 \times \frac{SN^2}{70{,}500}$$

where

L = length of rod string, feet
S = surface stroke, inches
N = pumping speed, strokes/min
E_{ot} = plunger overtravel, inches.

The term $SN^2/70{,}500$ in the revised equation is called the *acceleration factor* and is denoted by α.

Refer again to the example problem used to illustrate rod and tubing stretch. In this problem, $L = 5{,}000'$, $S = 54$ inches.

Assume that the well is being pumped at a speed of 20 strokes/minute (spm).

$$\alpha = \frac{SN^2}{70{,}500} = \frac{54 \times (20)^2}{70{,}500} = 0.306$$

$$E_{ot} = 1.55\left(\frac{L}{1{,}000}\right)^2$$

$$E_{ot} = 1.55\left(\frac{5{,}000}{1{,}050}\right)^2 \times 0.306 = 11.9''$$

From the previous calculations,

$$E_r = 10.3''$$
$$E_t = 30.9''$$

Now the actual plunger stroke, S_p, can be determined.

$$S_p = S - E_r - E_t + E_{ot}$$

where

E_r = rod stretch, inches
E_t = tubing stretch, inches
E_{ot} = plunger overtravel, inches
S = surface stroke, inches

For the example,

$$S_p = 54 - 10.3 - 30.9 + 11.9 = 24.7 \text{ inches.}$$

Obviously this amount of rod stretch is unacceptable. A different pump size, stroke length, and speed would be selected. The purpose of the example is to demonstrate the magnitude of stroke loss that can be encountered.

Pumping Unit Load Calculations

Pump volume calculations have been considered because the first step in a pumping unit design is to select a pump size, stroke length, and pumping speed to produce the desired volume of fluid. After this step has been taken, the pumping unit and prime mover size are determined. There are two general approaches to pumping unit load calculations. The method that was used exclusively until 1967 is commonly referred to as the Mills method. In 1967 the API published its RP 11L design method. Today the RP 11L method is more widely used. Neither method can be depended upon to give reliable predictions under all conditions. Actual loads should be monitored to be sure a unit is not operating in an overloaded condition. Since the design of a pumping installation by either the Mills or the API RP 11L method is a trial-and-error solution, many companies have developed mainframe computer programs to speed up the calculations and aid in optimizing the design. Programs are also available for programmable hand-held calculators and personal computers. The API RP 11L is used for most computer calculations. While computer-derived solutions save time in performing the calculations, they are not more accurate than hand calculations, so don't be mesmerized by a computer output printed out to ten significant figures; the results are not exact.

The first step in using either the Mills or the API method is to select a pump size and sucker rod string.

Pump Selection and Sucker Rod String Design

It was demonstrated in the discussion of sucker rod and tubing stretch that the selection of the optimum pump size is very important. The example problem illustrated that excessive rod stretch could occur if too large a plunger diameter was selected in an attempt to get a maximum producing rate.

Pump size selection. The optimum pump size for a given installation can be determined by trial and error if a computer design program is available. Table 12.2, which was published in 1958 in the *Bethlehem Sucker Rod Handbook,* is very helpful in selecting pump sizes for surface strokes of 74 inches or less. For example, the table shows that if it is desired to produce 500 BPD from a depth of 5,000 feet, then either a 2" or a 2¼" pump should be used.

Table 12.2 is a good starting point in picking the optimum pump size, although several sizes may have to be considered to find the optimum size. Since the table was published, pumping units with stroke lengths of 120 inches or more have become common.

Sucker rod string design. After the pump size has been selected for the first calculation attempt, the next step is to pick the rod sizes. Since most sucker rod strings of more than about 3,500 feet have more than one rod size to save unnecessary weight and to distribute the load better, the selection of rod sizes may appear to be a formidable task. Fortunately charts or tables are available to simplify this step.

There are two methods of designing a tapered sucker rod string. These two methods can best be understood by reference to figure 12.5. For illustrative purposes, a ¾" × ⅞" taper is shown.

The weight of the fluid is supported by all rods in the string. The rod that is connected to the pump has only the fluid load to support, while the rod at the top of the hole must support the fluid load plus the weight of all the rods below. The lower part of the rod string supports less load, so it can be smaller.

In the first method, the length of ¾" rods is determined by the length that will place the top rod at a stress equal to its working stress. In other words, the top rod of the smallest diameter is fully loaded. The second method, which is normally used, is to proportion the length of each rod size so that the top rod of

TABLE 12.2
PUMP PLUNGER SIZES RECOMMENDED FOR OPTIMUM CONDITIONS

Net Lift of Fluid (ft)	Fluid Production (bbl/day, 80% efficiency)									
	100	200	300	400	500	600	700	800	900	1000
2000	1½	1¾	2	2¼	2½	2¾	2¾	2¾	2¾	2¾
	1¼	1½	1¾	2	2¼	2½				
3000	1½	1¾	2	2¼	2½	2½	2¾	2¾	2¾	2¾
	1¼	1½	1¾	2	2¼	2¼	2½			
4000	1¼	1¾	2	2¼	2¼	2¼	2¼	2¼		
		1½	1¾	2	2					
5000	1¼	1¾	2	2	2¼	2¼				
		1½	1¾	1¾	2					
6000	1¼	1½	1¾	1¾						
		1¼	1½							
7000	1¼	1½								
	1⅛	1¼								
8000	1¼									
	1⅛									

In this tabulation, surface pumping strokes up to 74 in. only are considered.

SOURCE: *Sucker Rod Handbook* (Bethlehem, Pa.: Bethlehem Steel Company, 1958)

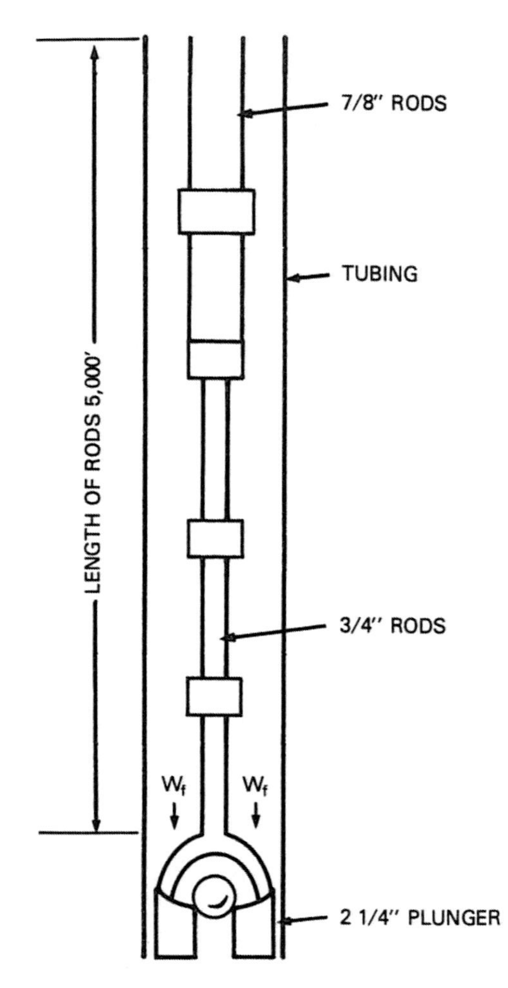

Figure 12.5. Tapered sucker rod string design

each size is under the same stress. Table 12.3 from API RP 11L lists the rod percentages to be used with different sizes of pumps in order to achieve equal stress in the top rod of each size.

In order to use the table, it is necessary to understand the way that API designates rod sizes. In the far left-hand column, the rod size is given as a two-digit number. Standard rods come in ⅛-inch increments. For example, common rod sizes are specified by their diameter as follows: ½", ⅝", ¾", ⅞", 1", and 1⅛".

The two-digit number in the table refers to the largest and the smallest rod size expressed in ⅛" increments. In other words, a *76* rod string is made up of ⅞" and ¾" rods. By using this information, the percentage of each rod size can be determined from the table. Since a 2¼" pump is used in figure 12.5, follow the rod number column down to 76. Then select the plunger diameter of 2.25 inches for a 76-rod string. Table 12.3 will show that for this combination, 46.5% of the rods should be ⅞" and 53.5% should be ¾".

Since the example rod string is 5,000 feet long,

Length of ¾" rods = 0.535 × 5,000 = 2,675 feet

Length of ⅞" rods = 0.465 × 5,000 = 2,325 feet.

TABLE 12.3
ROD AND PUMP DATA

Rod* No.	Plunger Diam., inches D	Rod Weight, lb/ft W_r	Elastic Constant, in./lb-ft E_r	Frequency Factor F_c	Rod String, % of each size 1⅛	1	⅞	¾	⅝	½
44	All	0.726	1.990×10^{-6}	1.000						100.0
54	1.06	0.908	1.668×10^{-6}	1.138					44.6	55.4
54	1.25	0.929	1.633×10^{-6}	1.140					49.5	50.5
54	1.50	0.957	1.584×10^{-6}	1.137					56.4	43.6
54	1.75	0.990	1.525×10^{-6}	1.122					64.6	35.4
54	2.00	1.027	1.460×10^{-6}	1.095					73.7	26.3
54	2.25	1.067	1.391×10^{-6}	1.061					83.4	16.6
54	2.50	1.108	1.318×10^{-6}	1.023					93.5	6.5
55	All	1.135	1.270×10^{-6}	1.000					100.0	
64	1.06	1.164	1.382×10^{-6}	1.229				33.3	33.1	33.5
64	1.25	1.211	1.319×10^{-6}	1.215				37.2	35.9	26.9
64	1.50	1.275	1.232×10^{-6}	1.184				42.3	40.4	17.3
64	1.75	1.341	1.141×10^{-6}	1.145				47.4	45.2	7.4

TABLE 12.3, *continued*

Rod* No.	Plunger Diam., inches D	Rod Weight, lb/ft W_r	Elastic Constant, in./lb-ft E_r	Frequency Factor F_c	Rod String, % of each size 1⅛	1	⅞	¾	⅝	½
65	1.06	1.307	1.138×10^{-6}	1.098				34.4	65.6	
65	1.25	1.321	1.127×10^{-6}	1.104				37.3	62.7	
65	1.50	1.343	1.110×10^{-6}	1.110				41.8	58.2	
65	1.75	1.369	1.090×10^{-6}	1.114				46.9	53.1	
65	2.00	1.394	1.070×10^{-6}	1.114				52.0	48.0	
65	2.25	1.426	1.045×10^{-6}	1.110				58.4	41.6	
65	2.50	1.460	1.018×10^{-6}	1.099				65.2	34.8	
65	2.75	1.497	0.990×10^{-6}	1.082				72.5	27.5	
65	3.25	1.574	0.930×10^{-6}	1.037				88.1	11.9	
66	All	1.634	0.883×10^{-6}	1.000				100.0		
75	1.06	1.566	0.997×10^{-6}	1.191			27.0	27.4	45.6	
75	1.25	1.604	0.973×10^{-6}	1.193			29.4	29.8	40.8	
75	1.50	1.664	0.935×10^{-6}	1.189			33.3	33.3	33.3	
75	1.75	1.732	0.892×10^{-6}	1.174			37.8	37.0	25.1	
75	2.00	1.803	0.847×10^{-6}	1.151			42.4	41.3	16.3	
75	2.25	1.875	0.801×10^{-6}	1.121			46.9	45.8	7.2	
76	1.06	1.802	0.816×10^{-6}	1.072			28.5	71.5		
76	1.25	1.814	0.812×10^{-6}	1.077			30.6	69.4		
76	1.50	1.833	0.804×10^{-6}	1.082			33.8	66.2		
76	1.75	1.855	0.795×10^{-6}	1.088			37.5	62.5		
76	2.00	1.880	0.785×10^{-6}	1.093			41.7	58.3		
76	2.25	1.908	0.774×10^{-6}	1.096			46.5	53.5		
76	2.50	1.934	0.764×10^{-6}	1.097			50.8	49.2		
76	2.75	1.967	0.751×10^{-6}	1.094			56.5	43.5		
76	3.25	2.039	0.722×10^{-6}	1.078			68.7	31.3		
76	3.75	2.119	0.690×10^{-6}	1.047			82.3	17.7		
77	All	2.224	0.649×10^{-6}	1.000			100.0			
85	1.06	1.883	0.873×10^{-6}	1.261		22.2	22.4	22.4	33.0	
85	1.25	1.943	0.841×10^{-6}	1.253		23.9	24.2	24.3	27.6	
85	1.50	2.039	0.791×10^{-6}	1.232		26.7	27.4	26.8	19.2	
85	1.75	2.138	0.738×10^{-6}	1.201		29.6	30.4	29.5	10.5	
86	1.06	2.058	0.742×10^{-6}	1.151		22.6	23.0	54.3		
86	1.25	2.087	0.732×10^{-6}	1.156		24.3	24.5	51.2		
86	1.50	2.133	0.717×10^{-6}	1.162		26.8	27.0	46.3		
86	1.75	2.185	0.699×10^{-6}	1.164		29.4	30.0	40.6		
86	2.00	2.247	0.679×10^{-6}	1.161		32.8	33.2	33.9		
86	2.25	2.315	0.656×10^{-6}	1.153		36.9	36.0	27.1		
86	2.50	2.385	0.633×10^{-6}	1.138		40.6	39.7	19.7		
86	2.75	2.455	0.610×10^{-6}	1.119		44.5	43.3	12.2		
87	1.06	2.390	0.612×10^{-6}	1.055		24.3	75.7			
87	1.25	2.399	0.610×10^{-6}	1.058		25.7	74.3			
87	1.50	2.413	0.607×10^{-6}	1.062		27.7	72.3			
87	1.75	2.430	0.603×10^{-6}	1.066		30.3	69.7			
87	2.00	2.450	0.598×10^{-6}	1.071		33.2	66.8			
87	2.25	2.472	0.594×10^{-6}	1.075		36.4	63.6			
87	2.50	2.496	0.588×10^{-6}	1.079		39.9	60.1			
87	2.75	2.523	0.582×10^{-6}	1.082		43.9	56.1			
87	3.25	2.575	0.570×10^{-6}	1.084		51.6	48.4			
87	3.75	2.641	0.556×10^{-6}	1.078		61.2	38.8			
87	4.75	2.793	0.522×10^{-6}	1.038		83.6	16.4			
88	All	2.904	0.497×10^{-6}	1.000		100.0				
96	1.06	2.382	0.670×10^{-6}	1.222	19.1	19.2	19.5	42.3		
96	1.25	2.435	0.655×10^{-6}	1.224	20.5	20.5	20.7	38.3		
96	1.50	2.511	0.633×10^{-6}	1.223	22.4	22.5	22.8	32.3		
96	1.75	2.607	0.606×10^{-6}	1.213	24.8	25.1	25.1	25.1		
96	2.00	2.703	0.578×10^{-6}	1.196	27.1	27.9	27.4	17.6		
96	2.25	2.806	0.549×10^{-6}	1.172	29.6	30.7	29.8	9.8		

TABLE 12.3, *continued*

Rod* No.	Plunger Diam., inches D	Rod Weight, lb/ft W_r	Elastic Constant, in./lb-ft E_r	Frequency Factor F_c	Rod String, % of each size					
					1⅛	1	⅞	¾	⅝	½
97	1.06	2.645	0.568×10^{-6}	1.120	19.6	20.0	60.3	......	......	......
97	1.25	2.670	0.563×10^{-6}	1.124	20.8	21.2	58.0	......	......	......
97	1.50	2.707	0.556×10^{-6}	1.131	22.5	23.0	54.5	......	......	......
97	1.75	2.751	0.548×10^{-6}	1.137	24.5	25.0	50.4	......	......	......
97	2.00	2.801	0.538×10^{-6}	1.141	26.8	27.4	45.7	......	......	......
97	2.25	2.856	0.528×10^{-6}	1.143	29.4	30.2	40.4	......	......	......
97	2.50	2.921	0.515×10^{-6}	1.141	32.5	33.1	34.4	......	......	......
97	2.75	2.989	0.503×10^{-6}	1.135	36.1	35.3	28.6	......	......	......
97	3.25	3.132	0.475×10^{-6}	1.111	42.9	41.9	15.2	......	......	......
98	1.06	3.068	0.475×10^{-6}	1.043	21.2	78.8	......	......	......	......
98	1.25	3.076	0.474×10^{-6}	1.045	22.2	77.8	......	......	......	......
98	1.50	3.089	0.472×10^{-6}	1.048	23.8	76.2	......	......	......	......
98	1.75	3.103	0.470×10^{-6}	1.051	25.7	74.3	......	......	......	......
98	2.00	3.118	0.468×10^{-6}	1.055	27.7	72.3	......	......	......	......
98	2.25	3.137	0.465×10^{-6}	1.058	30.1	69.9	......	......	......	......
98	2.50	3.157	0.463×10^{-6}	1.062	32.7	67.3	......	......	......	......
98	2.75	3.180	0.460×10^{-6}	1.066	35.6	64.4	......	......	......	......
98	3.25	3.231	0.453×10^{-6}	1.071	42.2	57.8	......	......	......	......
98	3.75	3.289	0.445×10^{-6}	1.074	49.7	50.3	......	......	......	......
98	4.75	3.412	0.428×10^{-6}	1.064	65.7	34.3	......	......	......	......
99	All	3.676	0.393×10^{-6}	1.000	100.0	......	......	......	......	......
107	1.06	2.977	0.524×10^{-6}	1.184	16.9	16.8	17.1	49.1	......	......
107	1.25	3.019	0.517×10^{-6}	1.189	17.9	17.8	18.0	46.3	......	......
107	1.50	3.085	0.506×10^{-6}	1.195	19.4	19.2	19.5	41.9	......	......
107	1.75	3.158	0.494×10^{-6}	1.197	21.0	21.0	21.2	36.9	......	......
107	2.00	3.238	0.480×10^{-6}	1.195	22.7	22.8	23.1	31.4	......	......
107	2.25	3.336	0.464×10^{-6}	1.187	25.0	25.0	25.0	25.0	......	......
107	2.50	3.435	0.447×10^{-6}	1.174	26.9	27.7	27.1	18.2	......	......
107	2.75	3.537	0.430×10^{-6}	1.156	29.1	30.2	29.3	11.3	......	......
108	1.06	3.325	0.447×10^{-6}	1.097	17.3	17.8	64.9	......	......	......
108	1.25	3.345	0.445×10^{-6}	1.101	18.1	18.6	63.2	......	......	......
108	1.50	3.376	0.441×10^{-6}	1.106	19.4	19.9	60.7	......	......	......
108	1.75	3.411	0.437×10^{-6}	1.111	20.9	21.4	57.7	......	......	......
108	2.00	3.452	0.432×10^{-6}	1.117	22.6	23.0	54.3	......	......	......
108	2.25	3.498	0.427×10^{-6}	1.121	24.5	25.0	50.5	......	......	......
108	2.50	3.548	0.421×10^{-6}	1.124	26.5	27.2	46.3	......	......	......
108	2.75	3.603	0.415×10^{-6}	1.126	28.7	29.6	41.6	......	......	......
108	3.25	3.731	0.400×10^{-6}	1.123	34.6	33.9	31.6	......	......	......
108	3.75	3.873	0.383×10^{-6}	1.108	40.6	39.5	19.9	......	......	......
109	1.06	3.839	0.378×10^{-6}	1.035	18.9	81.1	......	......	......	......
109	1.25	3.845	0.378×10^{-6}	1.036	19.6	80.4	......	......	......	......
109	1.50	3.855	0.377×10^{-6}	1.038	20.7	79.3	......	......	......	......
109	1.75	3.867	0.376×10^{-6}	1.040	22.1	77.9	......	......	......	......
109	2.00	3.880	0.375×10^{-6}	1.043	23.7	76.3	......	......	......	......
109	2.25	3.896	0.374×10^{-6}	1.046	25.4	74.6	......	......	......	......
109	2.50	3.911	0.372×10^{-6}	1.048	27.2	72.8	......	......	......	......
109	2.75	3.930	0.371×10^{-6}	1.051	29.4	70.6	......	......	......	......
109	3.25	3.971	0.367×10^{-6}	1.057	34.2	65.8	......	......	......	......
109	3.75	4.020	0.363×10^{-6}	1.063	39.9	60.1	......	......	......	......
109	4.75	4.120	0.354×10^{-6}	1.066	51.5	48.5	......	......	......	......
1010	All	4.538	0.318×10^{-6}	1.000	100.0	-----	-----	-----	-----	-----

*Rod No. shown in first column refers to the largest and smallest rod size in eighths of an inch. For example, Rod No. 76 is a two-way taper of ⅞ amd ⁶⁄₈ rods. Rod No. 85 is a four-way taper of ⁸⁄₈, ⅞, ⁶⁄₈, and ⅝ rods. Rod No. 109 is a two-way taper of 1¼ and 1⅛ rods. Rod No. 77 is a straight string of ⅞ rods, etc.

SOURCE: API RP 11L, *Design Calculations for Sucker Rod Pumping Systems (Conventional Units)*

Mills Method

The conventional design method used before the API RP 11L method was published is commonly called the Mills method after K. N. Mills, who suggested a method for calculating dynamic polished rod loads. The formula for the acceleration factor $\alpha = SN^2/70{,}500$ discussed in the overtravel calculations was proposed by Mills.

General. Pumping unit designs require a trial-and-error procedure. The first step is to decide on the amount of fluid that needs to be produced and the working fluid level that will be required to produce the desired quantity. After the fluid volume and working fluid level have been determined, the following equipment and pumping conditions are selected:

Pump size
Pumping speed (N)
Stroke length (S)
Rod string
Tubing size (anchored or not anchored)

The selection of a stroke length is based on the available stroke lengths for different-sized units. Therefore it is necessary to make a tentative selection of pumping unit size so that an appropriate stroke length can be selected. A tabulation of API pumping units showing the stroke lengths and load ratings is shown in table 12.4.

It would appear at first that it is difficult to make all of the assumptions necessary to perform pumping unit calculations. Obviously, with experience it is possible to make more intelligent selections and assumptions, but it is not too difficult even for a novice.

Production. After the pump, stroke length, and speed have been selected and a rod string has been decided upon, the production can be calculated. The method outlined in the preceding section should be followed to allow for rod and tubing stretch and plunger overtravel. The volume determined by these calculations gives the theoretical pump displacement. It is necessary to multiply the theoretical pump displacement by the *pump efficiency, Ev,* to get the estimated producing rates. Pump efficiency accounts for fluid slippage past the plunger and any entrainment of gas in the fluid. If the well is producing with a high water cut, the pump efficiency could be 90%. If gassy fluids are being pumped, *Ev* may be 60% or lower. An *Ev* of 80% to 85% is probably a good estimate when no data are available. The estimation of *Ev* should always be adjusted according to local experience.

Polished rod load. A typical rod-pump system (fig. 12.5) shows that the static load imposed on the pumping unit is composed of the weight of the rods plus the weight of the fluid. When the well is pumping, an additional dynamic load is imposed. Mills accounted for this additional dynamic component by the acceleration factor α previously discussed. The peak polished rod load (*PPRL*) is

$$PPRL = W_r + W_f + W_r\alpha$$

where

$PPRL$ = peak polished rod load, lb
W_r = weight of rods, lb
W_f = weight of fluid, lb
α = Mills acceleration factor
$\alpha = \dfrac{SN^2}{70{,}500}$

The equation can be simplified as follows:

$$PPRL = W_f + W_r\,(1 + \alpha)$$

The term $1 + \alpha$ is commonly called the *impulse factor.*

It is also important to determine the minimum polished rod load, *MPRL.* The minimum polished rod load occurs at the top of the stroke when the rods tend to continue in an upward direction and the acceleration factor decreases the polished rod load. In calculating the *PPRL,* Mills assumes that friction losses will offset the buoyancy of the rods. On the downstroke, Mills accounts for the buoyant weight of the rods.

$$MPRL = W_r - W_r\alpha - \frac{62.4G\,W_f}{490}$$

TABLE 12.4
PUMPING UNIT SIZE RATINGS

Pumping Unit Size	Reducer Rating, in.-lb	Structure Capacity, lb	Max. Stroke Length, in.
6.4— 32— 16	6,400	3,200	16
6.4— 21— 24	6,400	2,100	24
10— 32— 24	10,000	3,200	24
10— 40— 20	10,000	4,000	20
16— 27— 30	16,000	2,700	30
16— 53— 30	16,000	5,300	30
25— 53— 30	25,000	5,300	30
25— 56— 36	25,000	5,600	36
25— 67— 36	25,000	6,700	36
40— 89— 36	40,000	8,900	36
40— 76— 42	40,000	7,600	42
40— 89— 42	40,000	8,900	42
40— 76— 48	40,000	7,600	48
57— 76— 42	57,000	7,600	42
57— 89— 42	57,000	8,900	42
57— 95— 48	57,000	9,500	48
57—109— 48	57,000	10,900	48
57— 76— 54	57,000	7,600	54
80—109— 48	80,000	10,900	48
80—133— 48	80,000	13,300	48
80—119— 54	80,000	11,900	54
80—133— 54	80,000	13,300	54
80—119— 64	80,000	11,900	64
114—133— 54	114,000	13,300	54
114—143— 64	114,000	14,300	64
114—173— 64	114,000	17,300	64
114—143— 74	114,000	14,300	74
114—119— 86	114,000	11,900	86
160—173— 64	160,000	17,300	64
160—143— 74	160,000	14,300	74
160—173— 74	160,000	17,300	74
160—200— 74	160,000	20,000	74
160—173— 86	160,000	17,300	86
228—173— 74	228,000	17,300	74
228—200— 74	228,000	20,000	74
228—213— 86	228,000	21,300	86
228—246— 86	228,000	24,600	86
228—173—100	228,000	17,300	100
228—213—120	228,000	21,300	120
320—213— 86	320,000	21,300	86
320—256—100	320,000	25,600	100
320—305—100	320,000	30,500	100
320—213—120	320,000	21,300	120
320—256—120	320,000	25,600	120
320—256—144	320,000	25,600	144
456—256—120	456,000	25,600	120
456—305—120	456,000	30,500	120
456—365—120	456,000	36,500	120
456—256—144	456,000	25,600	144
456—305—144	456,000	30,500	144
456—305—168	456,000	30,500	168
640—305—120	640,000	30,500	120
640—256—144	640,000	25.600	144
640—305—144	640,000	30,500	144
640—365—144	640,000	36,500	144
640—305—168	640,000	30,500	168
640—305—192	640,000	30,500	192
912—427—144	912,000	42,700	144
912—305—168	912,000	30,500	168
912—365—168	912,000	36,500	168
912—305—192	912,000	30,500	192
912—427—192	912,000	42,700	192
912—470—240	912,000	47,000	240
912—427—216	912,000	42,700	216
1280—427—168	1,280,000	42,700	168
1280—427—192	1,280,000	42,700	192
1280—427—216	1,280,000	42,700	216
1280—470—240	1,280,000	47,000	240
1280—470—300	1,280,000	47,000	300
1824—427—192	1,824,000	42,700	192
1824—427—216	1,824,000	42,700	216
1824—470—240	1,824,000	47,000	240
1824—470—300	1,824,000	47,000	300
2560—470—240	2,560,000	47,000	240
2560—470—300	2,560,000	47,000	300
3648—470—240	3,648,000	47,000	240
3648—470—300	3,648,000	47,000	300

SOURCE: API Spec 11E, *Pumping Units,* 13th ed., August 1, 1984

The term $62.4G\ W_r/490$ accounts for the buoyancy of the rods. The buoyancy factor is determined by the weight of the fluid displaced by the rods. Since the density of steel is 490 lb/cu ft, $W_r/490$ gives the cubic feet of fluid displaced. Water weighs 62.4 lb/cu ft, and G, the specific gravity of the fluid, is used to correct the weight of water to the actual fluid being pumped. The *MPRL* equation can be simplified to

$$MPRL = W_r\,(1 - \alpha - 0.127G)$$

where

G = specific gravity of fluid being pumped.

Counterbalance effect. It is generally assumed that sufficient counterbalance is added to balance the average of the *PPRL* and *MPRL*.

Accounting for friction and buoyancy of the rods,

$$PPRL = W_f + W_r + W_r\alpha - \text{buoyancy} + \text{friction}$$

$$MPRL = W_r - W_r\alpha - \text{buoyancy} - \text{friction}$$

Since average load = $(PPRL + MPRL)/2$, counterbalance effect (CBE) is–

$$CBE = \frac{2W_r + W_f - 2 \times \text{buoyancy}}{2}$$

$$CBE = 0.5\ W_f + W_r - \text{buoyancy}$$

Since buoyancy = $0.127\ G\ W_r$,

$$CBE = 0.5\ W_f + W_r (1 - 0.127\ G).$$

Peak torque. Torque is the twisting force that needs to be applied to the gear box in order to pump a well at the desired speed. Torque is measured in inches/lb. Peak torque (PT) is estimated by the following formula:

$$PT_{up} = (PPRL - CBE)S/2$$

This equation gives peak torque on the upstroke. On the downstroke, peak torque is–

$$PT_{down} = (CBE - MPRL)S/2$$

Horsepower. The theoretical or hydraulic horsepower (HHP) required to lift a volume of fluid from a given depth can be calculated as follows:

1 horsepower (HP) = 33,000 ft-lb/min

$$HHP = \frac{L \times BPD \times 42 \times 8.33 \times 6}{1{,}440 \text{ min/day}} \times \frac{1}{33{,}000 \text{ ft-lb/min/HP}}$$

$$HHP = 7.36 \times 10^{-6} \times GL \times BPD$$

where

L = depth of lift, feet

G = fluid specific gravity.

HHP is the theoretical amount of horsepower needed to lift the fluid. Because of losses in the sucker rod string and the surface equipment, it has been common practice to multiply the hydraulic horsepower by a factor of 2 to 2.5 to obtain the size of prime mover actually needed.

One major manufacturer of pumping units suggests the following formulas:

For slow-speed engines and high-slip electric motors–

$$BHP = \frac{\text{depth (ft)} \times BPD}{56{,}000}$$

where

BHP = brake horsepower.

For multicylinder and normal-slip electric motors–

$$BHP = \frac{\text{depth (ft)} \times BPD}{45{,}000}$$

API RP 11L Method

The Mills method does not yield precise results when compared with measured values on actual installations. In order to overcome the problems with the Mills method, the API published its RP 11L design method in 1967. Although the RP 11L method gives better results than the Mills method, it still does not give precise information, and experience is still a factor in designing pumping unit installations.

General. The API RP 11L method was developed using an analog computer. Charts were developed utilizing nondimensional parameters to calculate production, $PPRL$, peak torque, horsepower, and counterbalance effect. To simplify the use of RP 11L, the API published a form shown in figure 12.6. The minimum amount of information that must be available to evaluate a beam pumping installation is listed as follows:

Fluid level–H, the net lift in feet

Pump depth–L, feet

Company ________ Well Name ________ Date ________
Field ________ County ________ State ________
Required pump displacement, PD ________ bbls./day Maximum allowable rod stress ________ psi
Fluid level, D = ________ ft. Pumping speed, N = ________ SPM Plunger diameter, D_p = ________ in.
Pump depth, L = ________ ft. Length of stroke, S = ________ in. Spec. grav. of fluid, G = ________
Tubing size ________ in. Is it anchored? Yes, No Sucker rods ________

Record factors

1. W_r = ________ lb/ft
2. E_r = ________
3. F_c = ________
4. E_t = ________

Calculate non-dimensional variables:

5. $F_o = .340 \times G \times D_p^2 \times D = .340 \times$ ________ $\times$ ________ $\times$ ________ = ________ lbs.
6. $1/k_r = E_r \times L$ = ________ $\times$ ________ = ________ in/lb.
7. $Sk_r = S \div 1/k_r$ = ________ $\div$ ________ = ________ lbs.
8. F_o/Sk_r = ________ $\div$ ________ = ________
9. $N/N_o = NL \div 245{,}000$ = ________ $\times$ ________ $\div$ 245,000 = ________
10. $N/N_o' = N/N_o \div F_c$ = ________ $\div$ ________ = ________
11. $1/k_t = E_t \times L$ = ________ $\times$ ________ = ________ in/lb.

Solve for S_p and PD:

12. S_p/S = ________
13. $S_p = [(S_p/S) \times S] - [F_o \times 1/k_t]$ = [________ $\times$ ________] − [________ $\times$ ________] = ________ in.
14. $PD = 0.1166 \times S_p \times N \times D_p^2 = 0.1166 \times$ ________ $\times$ ________ $\times$ ________ = ________ barrels per day

If calculated pump displacement is unsatisfactory make appropriate adjustments in assumed data and repeat steps 1 through 14.

Determine non-dimensional parameters:

15. $W = W_r \times L$ = ________ $\times$ ________ = ________ lbs.
16. $W_{rf} = W[1 - (.128G)]$ = ________ [1 − (.128 $\times$ ________)] = ________ lbs.
17. W_{rf}/Sk_r = ________ $\div$ ________ = ________

Record non-dimensional factors

18. F_1/Sk_r = ________
19. F_2/Sk_r = ________
20. $2T/S^2k_r$ = ________
21. F_3/Sk_r = ________
22. T_a = ________

Solve for operating characteristics:

23. $PPRL = W_{rf} + [(F_1/Sk_r) \times Sk_r]$ = ________ + [________ $\times$ ________] = ________ lbs.
24. $MPRL = W_{rf} - [(F_2/Sk_r) \times Sk_r]$ = ________ − [________ $\times$ ________] = ________ lbs.
25. $PT = (2T/S^2k_r) \times Sk_r \times S/2 \times T_a$ = ________ $\times$ ________ $\times$ ________ $\times$ ________ = ________ in.-lbs.
26. $PRHP = (F_3/Sk_r) \times Sk_r \times S \times N \times 2.53 \times 10^{-6}$ = ________ $\times$ ________ $\times$ ________ $\times$ ________ $\times 2.53 \times 10^{-6}$ = ________
27. $CBE = 1.06\,(W_{rf} + \frac{1}{2} F_o) = 1.06 \times$ (________ + ________) = ________ lbs.

Remarks ________

Figure 12.6. API form for design calculations for conventional sucker rod pumping systems. From API RP 11L, courtesy of API.

Pumping speed–N, strokes per minute

Length of surface stroke–S, inches

Pump plunger diameter–D, inches

Specific gravity of fluid–G

Nominal tubing diameter and whether it is anchored

Sucker rod sizes and lengths, maximum allowable rod stress.

The above information must be known or assumed in order to evaluate a pumping installation by either the Mills or the API RP 11L method. The selection of a surface stroke length requires that at least a tentative beam unit selection be made.

The top portion of figure 12.6 has blanks for the known or assumed data discussed above. Blanks no. 1 through no. 4 are filled in with data

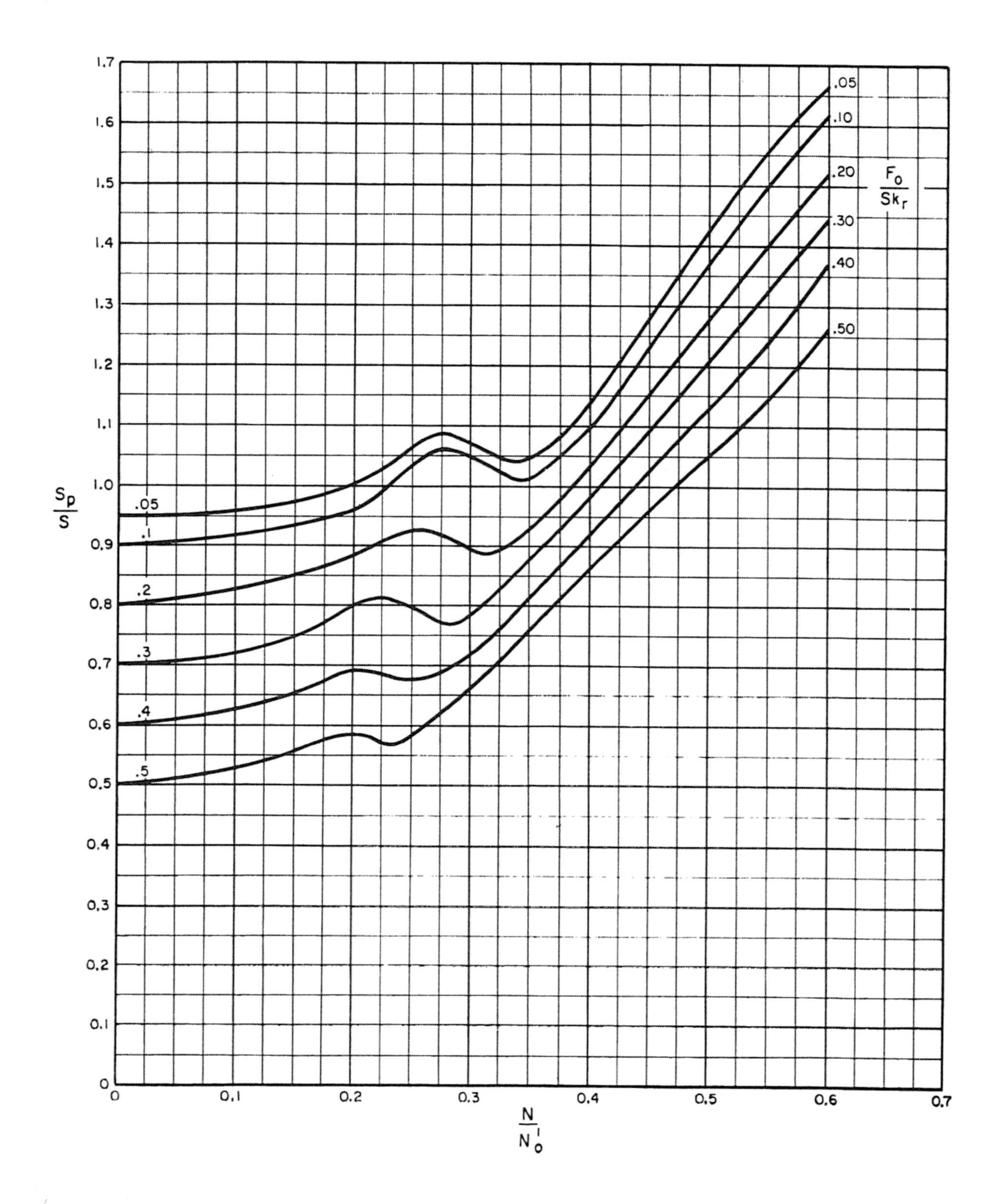

Figure 12.7. S_p/S, plunger stroke factor. From API RP 11L, courtesy of API.

from the sucker rod, pump, and tubing data in table 12.3 and table 12.5. Table 12.3 has been previously discussed in the sucker rod design section. Blanks no. 1 through no. 4 contain the following data:

1. W_r = weight of rods, lb/ft
2. E_r = elastic constant of rods, in./lb-ft
3. F_c = frequency factor for rods
4. E_t = elastic constant of tubing, inches/lb-ft. This value is obtained from table 12.5.

The next step after recording the above data is to calculate certain nondimensional variables in blanks no. 5 through no. 11.

5. F_o = fluid load in lb. The fluid load is based on the full plunger area.
6. $1/k_r$ = $E_r \times L$, inches/lb. This is the amount the rods will stretch for each pound of fluid load imposed.
7. Sk_r = $S \div 1/k_r$, pounds. This number indicates how many pounds are required to stretch the rod string the length of the surface stroke.
8. F_o/Sk_r = lb/lb.
9. N/N_o = NL ÷ 245,000
   where N = pumping speed, spm
   N_o = spm at natural frequency of single rod string
   L = length of rods.
10. N/N_o' = $N/N_o \div F_c$
    where N_o' = spm at natural frequency of a tapered rod string
11. $1/k_t$ = $E_t \times L$ This is the amount the tubing will stretch (in inches) for each pound of fluid load imposed.

After the above nondimensional variables have been calculated, the next step is to calculate pump displacement and estimate the producing rate.

Production. The pump displacement is calculated using the chart in figure 12.7. Note that this chart is a plot of the plunger stroke factor, S_p/S vs. N/N_o' for different values of F_o/Sk_r. Using the value of F_o/Sk_r from blank no. 8 and the value of N/N_o' obtained in blank no. 10, the stroke factor S_p/S is obtained. The stroke factor multiplied by the surface stroke gives the plunger stroke adjusted for rod stretch and plunger overtravel. In blank no. 13 the plunger stroke is calculated. Note that $S_p = [(S_p/S) \times S] - [F_o \times 1/k_t]$. The value $F_o \times 1/k_t$ is the tubing stretch. If the tubing is anchored, then this $F_o \times 1/k_t$ value is zero. The pump displacement is calculated in blank no. 14. If the pump displacement multiplied by a suitable pump efficiency factor results in the desired producing rate, then the load is calculated. If the producing rate is not enough or is a much larger value than desired, then some of the assumptions at the top of the form are revised and a new value of production is calculated.

Polished rod load. The next step is to calculate the peak polished rod load (*PPRL*) and the minimum polished rod load (*MPRL*).

The peak polished rod load is calculated using a value of F_1/Sk_r obtained from the chart in figure 12.8. Note that F_1/Sk_r is plotted against N/N_o for different values of F_o/Sk_r. The value of F_o/Sk_r in blank no. 8 and the value of N/N_o in blank no. 9 are used to obtain a value of F_1/Sk_r. The peak polished rod load is calculated in blank no. 23, using the F_1/Sk_r value from figure 12.8 and W_{rf}. W_{rf} is the buoyant weight of the rod string in pounds, calculated in blanks nos. 15 and 16.

The *MPRL* is calculated in a similar manner, using a value of F_2/Sk_r obtained from the chart in figure 12.9.

Counterbalance effect. The counterbalance effect (*CBE*) is calculated in blank no. 27, using the following formula:

$$CBE = 1.06\,(W_{rf} + \tfrac{1}{2}F_o)$$

where

W_{rf} = weight of rods in fluid
F_o = fluid load.

This equation is the same as the one used in the Mills method except that the constant 1.06 has been added.

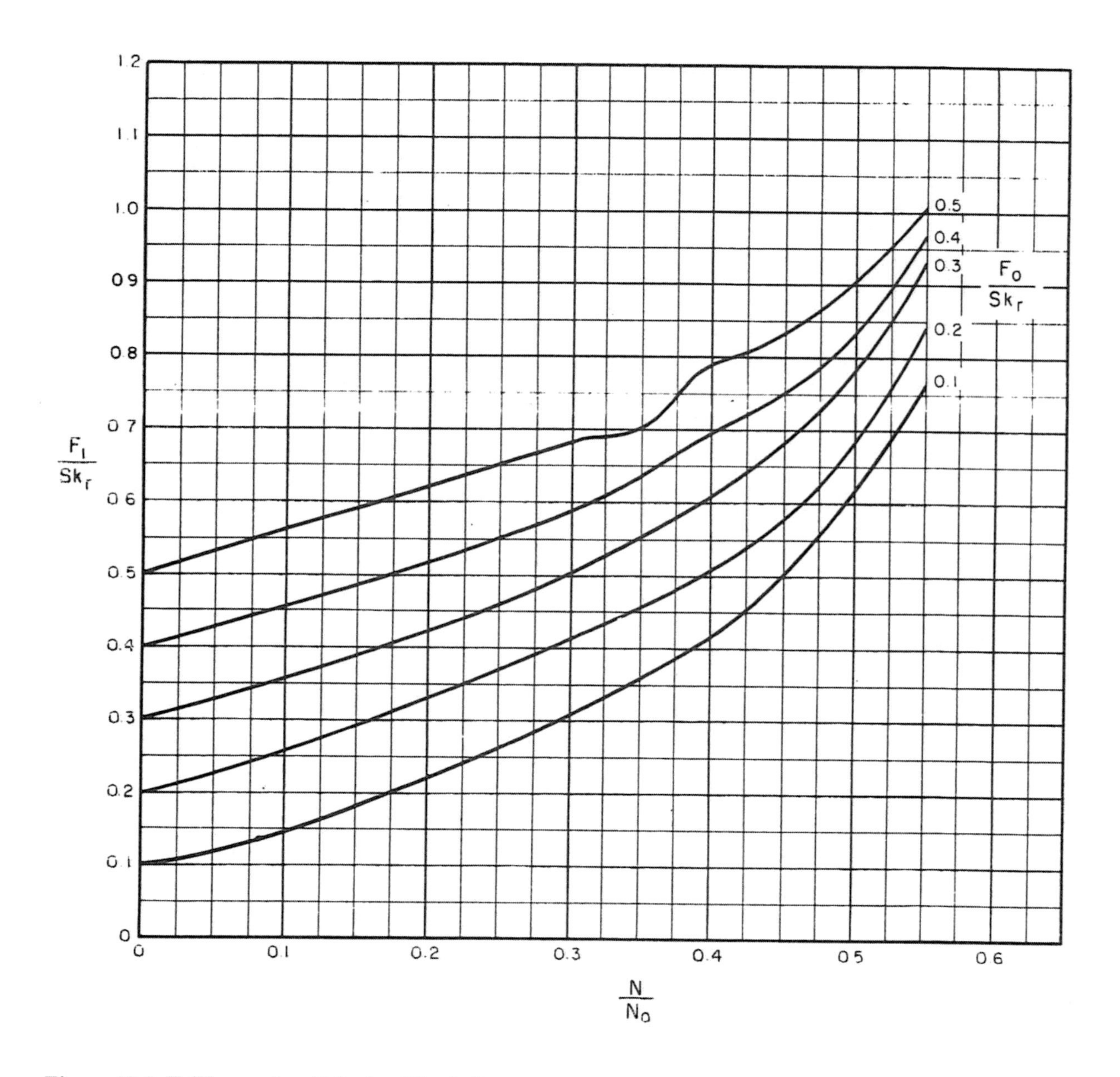

Figure 12.8. F_1/Sk_r, peak polished rod load. From API RP 11L, courtesy of API.

Peak torque. The peak torque (*PT*) is calculated in blank no. 25 using a value of $2T/S^2k_r$ obtained from the chart in figure 12.10. This chart is similar to the charts in figures 12.8 and 12.9. It is also necessary to obtain a torque adjustment factor, T_a, from the chart in figure 12.11. In order to calculate the peak torque, the nondimensional parameter W_{rf}/Sk_r, calculated in blank no. 17, is used.

Polished rod horsepower. The next step is to calculate the polished rod horsepower (*PRHP*) in blank no. 26. This is accomplished by using a value of F_3/Sk_r from the chart in figure 12.12. The other values for calculating *PRHP* are Sk_r, *S*, and *N*, which are available in the upper sections.

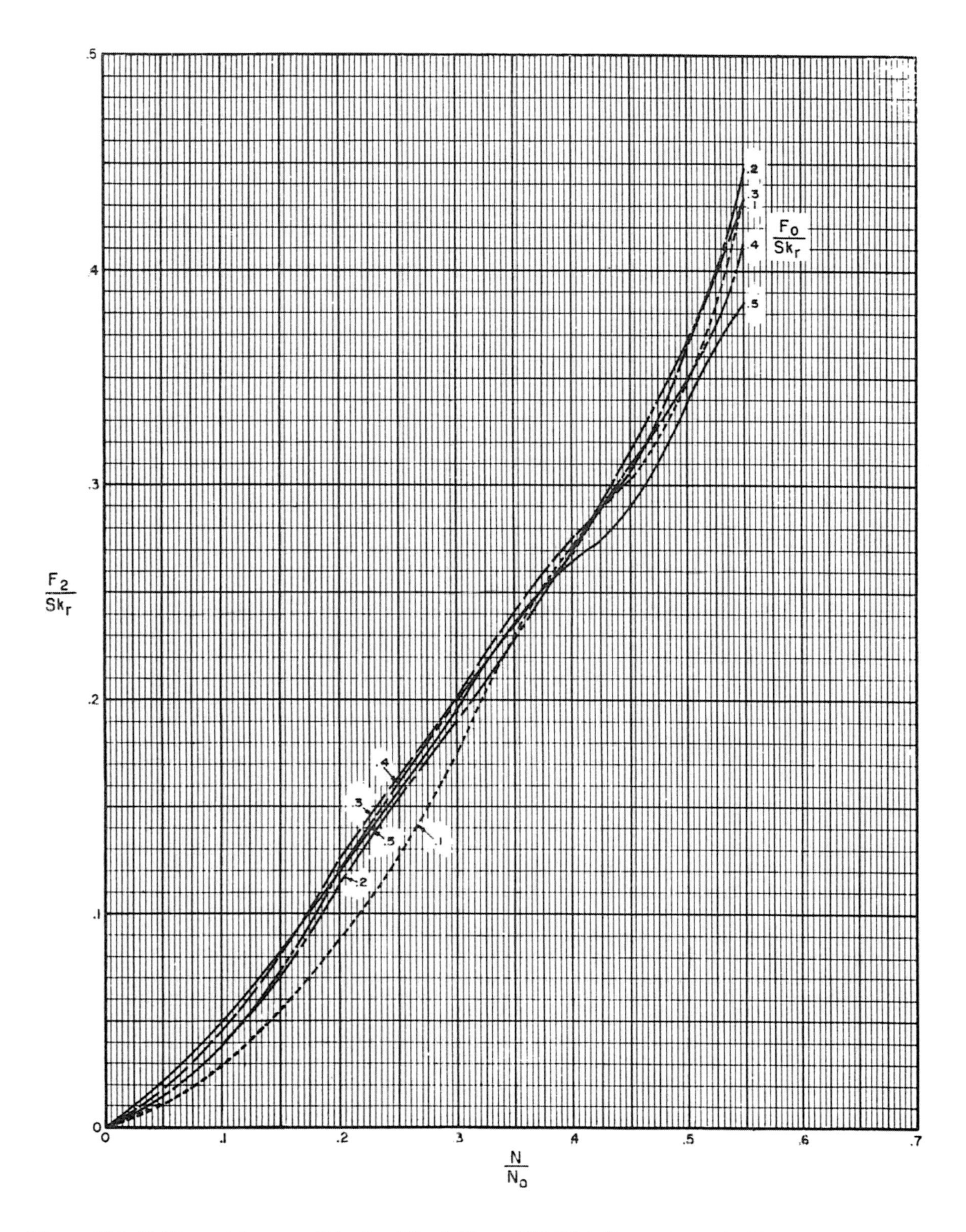

Figure 12.9. F_2/Sk_r, minimum polished rod load. From API RP 11L, courtesy of API.

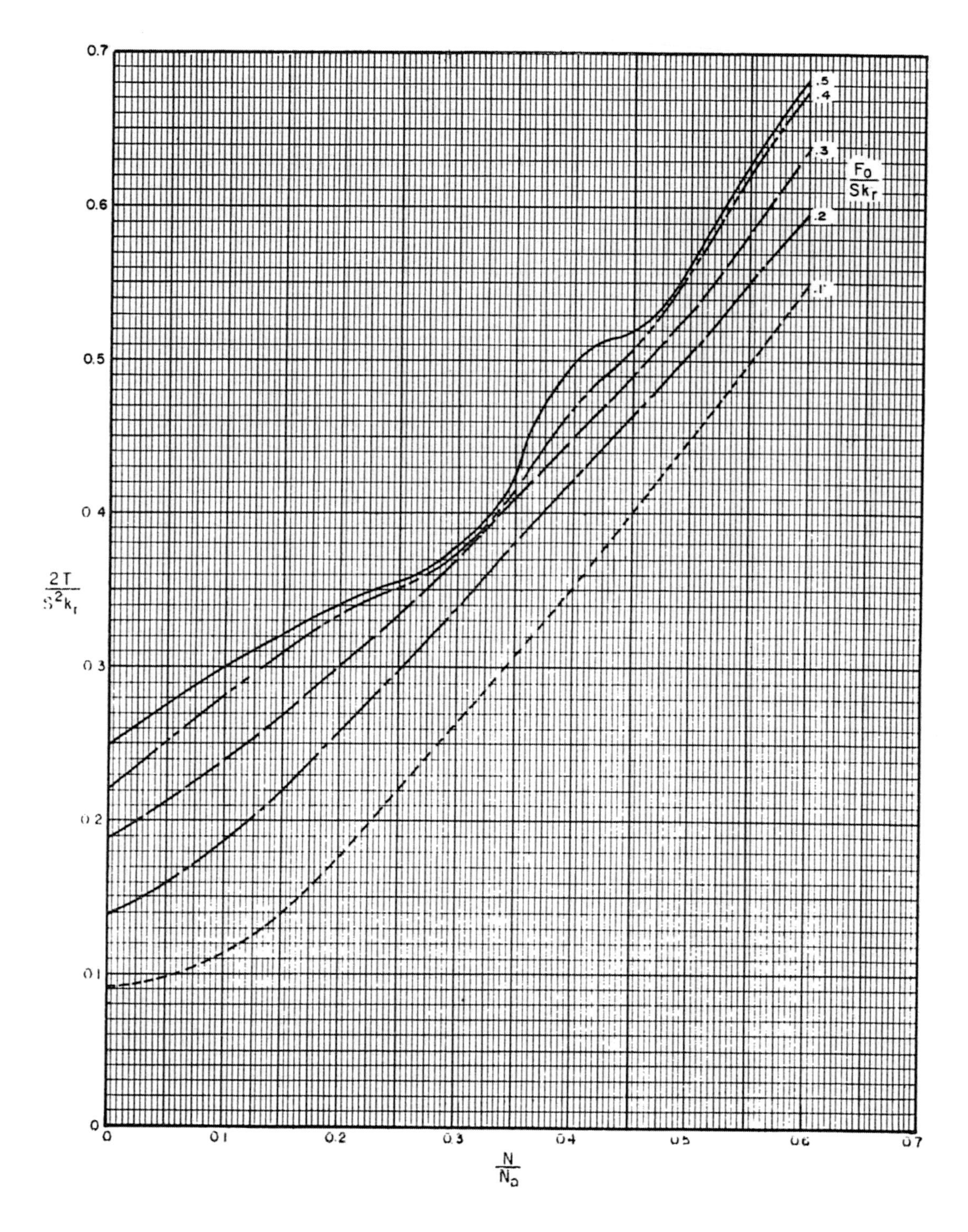

Figure 12.10. $2T/S^2k_r$, peak torque. From API RP 11L, courtesy of API.

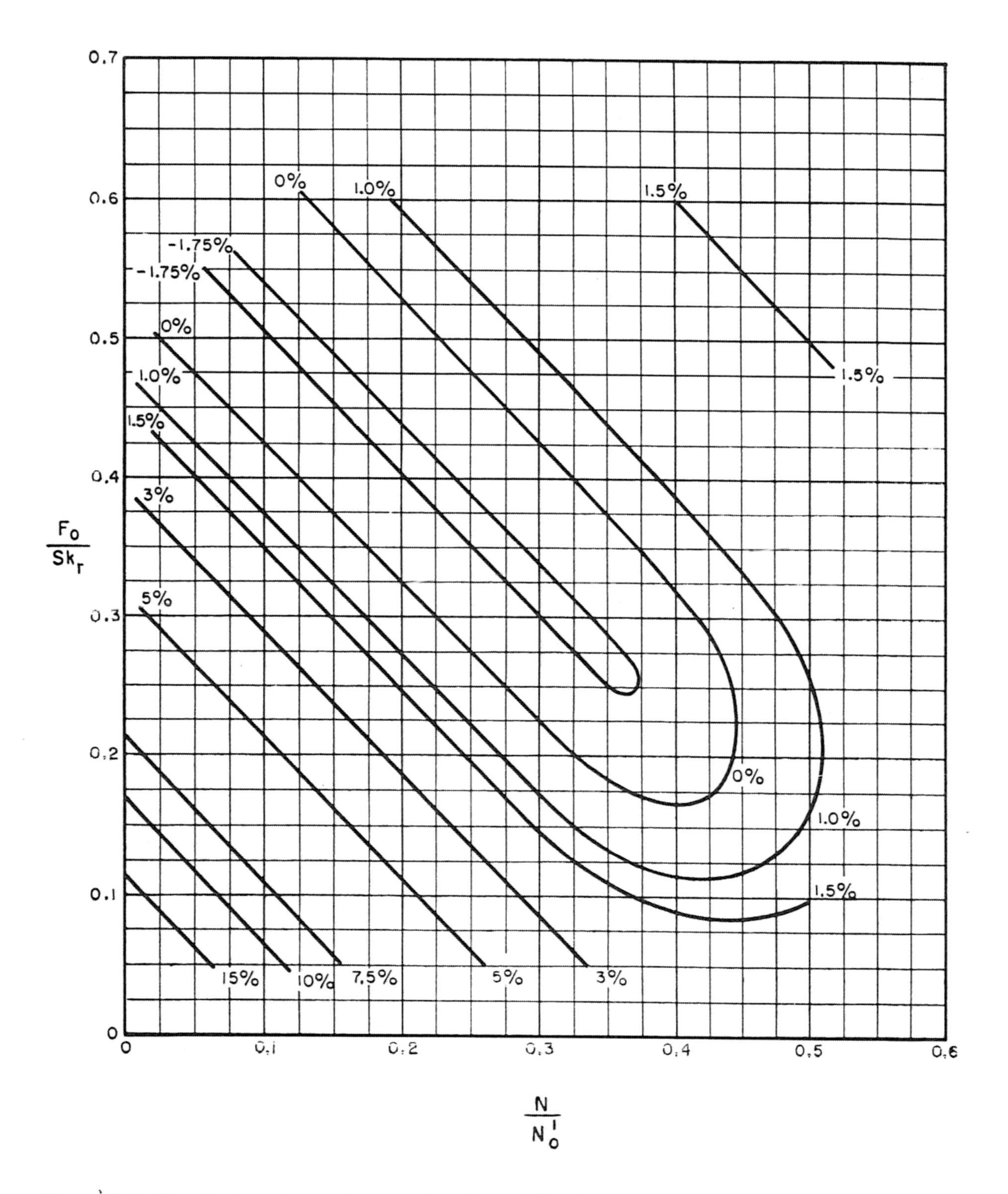

Figure 12.11. T_a, adjustment for peak torque for values of W_{rf}/Sk_r other than 0.3. From API RP 11L, courtesy of API.

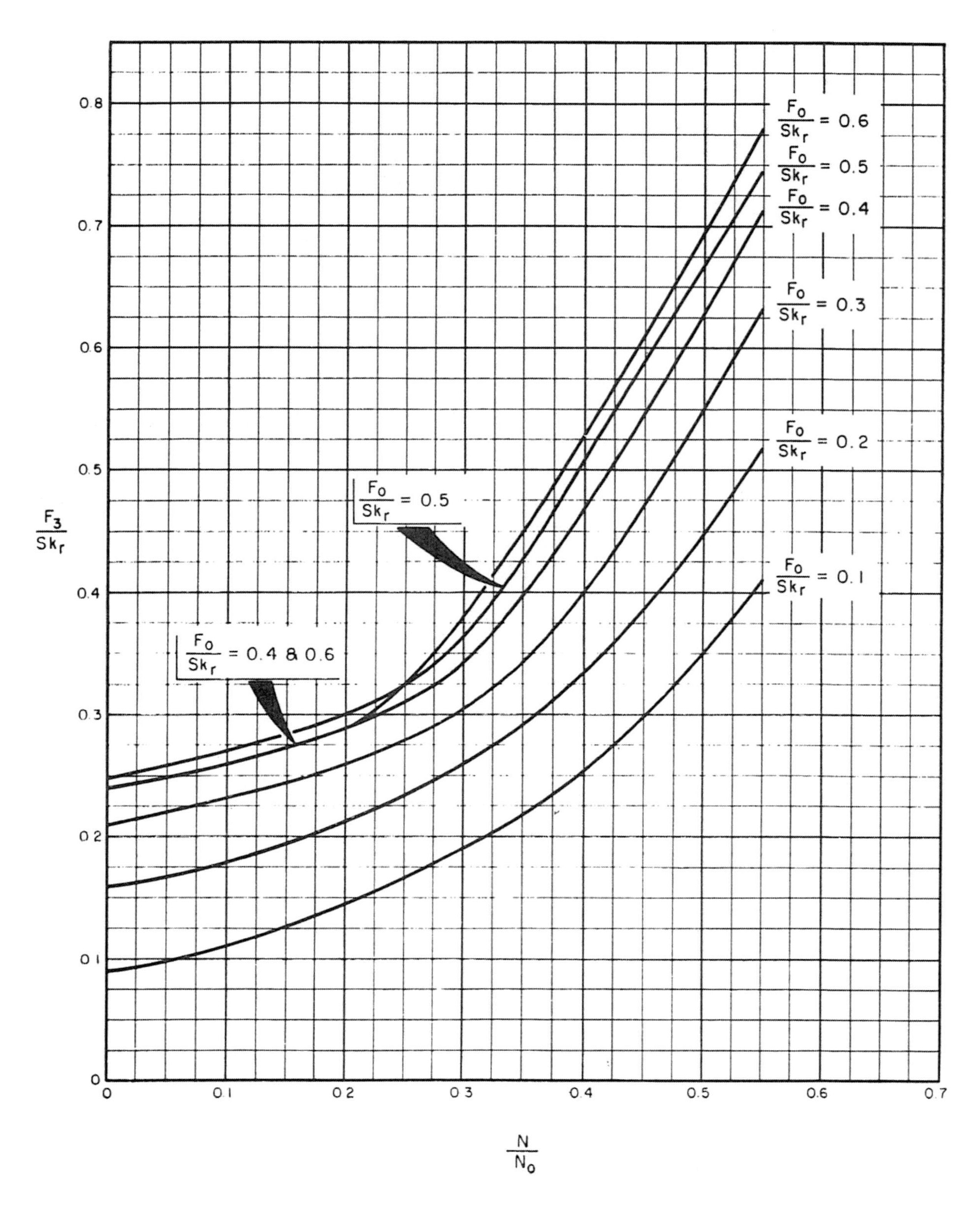

Figure 12.12. F_3/Sk_r, polished rod horsepower. From API RP 11L, courtesy of API.

Example Problem

The use of the Mills and API RP 11L methods can best be illustrated by means of an example problem. The API RP 11L brochure contains example design calculations. The data for this example will be used for both the Mills method and the API RP 11L method calculations. The results of the two methods will be compared. Following are the known or assumed data for this problem:

Fluid level, $H = 4{,}500$ ft
Pump depth, $L = 5{,}000$ ft
Tubing size = 2 ³⁄₈" OD, 4.6 lb/ft, not anchored
Pumping speed, $N = 16$ spm
Length of stroke, $S = 54$"
Plunger diameter, $D = 1.5$"
Specific gravity of fluid, $G = 0.9$
Sucker rods 33.8% of ⁷⁄₈", 66.2% of ³⁄₄"
Desired producing rate = 150 BPD
Pump volumetric efficiency = 85% (Assumed. No value given in RP 11L example problem.)

⁷⁄₈" sucker rods = 0.338 × 5,000 = 1,690 ft

$$A_r = \frac{\pi D^2}{4} = \frac{\pi \times 0.875^2}{4} = 0.601 \text{ inches}^2$$

³⁄₄" sucker rods = 0.662 × 5,000 = 3,310 ft

$$A_r = \frac{\pi \times 0.75^2}{4} = 0.442 \text{ inches}^2$$

The well will be pumped with an electric motor.

The fluid level, fluid specific gravity, producing rate, and pump depth are based on the well characteristics, so these are known data. The other data are assumed. The pump diameter can be estimated by referring to table 12.2. For a producing rate of 150 BPD from 4,500', a 1½" diameter is indicated. Since a 1½" pump is selected, then 2³⁄₈" OD tubing is the logical choice. The tubing is not anchored. The length of stroke, 54 inches, and the pump speed, $N = 16$ spm, are chosen on the basis of experience. Since the pump will be set below 3,500 feet, a tapered rod string is selected. From table 12.3 for a 1½" pump, a tapered string consisting of 33.8% ⁷⁄₈" rods and 66.2% ³⁄₄" rods is selected.

The volumetric efficiency of 85% is estimated on the basis of experience.

Mills Solution

The first step using the Mills method is to calculate the net plunger stroke S_p.

$$S_p = S - E_r - E_t + E_{ot}$$

where

E_r = rod stretch, inches
E_t = tubing stretch, inches
E_{ot} = plunger overtravel, inches
S = surface stroke, inches.

Since only the surface stroke S is known, E_r, E_t, and E_{ot} must be calculated. For a tapered rod string–

$$E_r = \frac{12W_f}{E} + \left(\frac{L_1}{A_{r1}} + \frac{L_2}{A_{r2}}\right)$$

where

W_f = weight of fluid, lb
A_{r1} and A_{r2} = rods' cross-sectional area, inches²
L_1 and L_2 = length of rods
$E = 30 \times 10^6$.

$$E_t = \frac{12W_f \times L}{E \times A_t}$$

where

A_t = tubing cross-sectional area, inches².

$$E_{ot} = 1.55\,\frac{(L)^2}{1{,}000}\,\alpha$$

where

$$\alpha = \frac{SN^2}{70{,}500}$$

$$W_f = 0.433\, G\, A_p$$

where

G = specific gravity of fluid
A_p = area of plunger, inches²
H = working fluid level.

The above formula for calculating the weight of the fluid, W_f, assumes that the fluid load acts on the full area of the plunger, A_p. This is the method used by the API in RP 11L. In the original Mills method the fluid load was assumed to act on the area of the plunger minus the area of the connected rod ($A_p - A_r$). Mills also assumed the working fluid level to be at the pump. Since this is usually not the case, the Mills method has been modified to calculate the fluid load based on the full plunger area and the actual working fluid level.

For a 1.5" plunger,

$$A_p = \frac{\pi \times 1.5^2}{4} = 1.767 \text{ in.}^2$$

$$H = 4{,}500 \text{ ft.}$$

$$W_f = 0.433 \times 0.9 \times 1.767 \times 4{,}500 \text{ (Water} = 0.433 \text{ psi/ft of height.)}$$

$$W_f = 3{,}098 \text{ lb.}$$

$$E_r = \frac{12W_f}{E} \times \left(\frac{L_1}{A_{r1}} + \frac{L_2}{A_{r2}}\right)$$

$$= \frac{12 \times 3{,}098}{30 \times 10^6}\left(\frac{1{,}690}{0.601} + \frac{3{,}310}{0.442}\right)$$

$$E_r = 12.8"$$

$$E_t = \frac{12W_f \times L}{E\,A_t}$$

For 2⅜" OD tubing (OD = 2.375", ID = 1.995"),

$$A_t = \frac{\pi}{4}(2.375^2 - 1.995^2) = 1.304 \text{ inches}^2$$

$$E_t = \frac{12 \times 3{,}098 \times 5{,}000}{30 \times 10^6 \times 1.304} = 4.8".$$

$$E_{ot} = 1.5\left(\frac{L}{1{,}000}\right)^2 \alpha$$

where

$$\alpha = \frac{SN^2}{70{,}500}$$

$$\alpha = \frac{54 \times 16^2}{70{,}500} = 0.196.$$

$$E_{ot} = 1.5\left(\frac{5{,}000}{1{,}000}\right)^2 \times 0.196$$

$$E_{ot} = 7.4".$$

$$S_p = S - E_r - E_t + E_{ot}$$

$$S_p = 54 - 12.8 - 4.8 + 7.4 = 43.8".$$

$$PD = KS_p N \quad (K = 0.262 \text{ from table 12.1})$$

$$PD = 0.262 \times 43.8 \times 16 = 184 \text{ BPD.}$$

Estimated production = $PD \times E_v$ (E_v = 85% assumed)

Estimated production = 184 × 0.85 = 156 BPD.

Since the desired producing rate is 150 BPD, the assumed data are satisfactory. If the production were less than desired, it would be necessary to recalculate using a different speed, stroke length, or pump size to achieve the desired rate. Each design then is a trial-and-error solution.

The next step is to calculate the *PPRL* and the *MPRL*.

$$PPRL = W_f + W_r(1 + \alpha)$$

where

α = 0.196 from overtravel calculation

W_r = 1.833 lb/ft, table 12.3 (Average weight for ¾" and ⅞" rods)

$$PPRL = 3{,}098 + 1.833 \times 5{,}000\,(1 + 0.196)$$

$$PPRL = 14{,}060 \text{ lb.}$$

$$MPRL = W_r(1 - \alpha - 0.127G) = 1.833 \times 5{,}000\,(1 - 0.196 - 0.127 \times 0.9)$$

$$MPRL = 6{,}320 \text{ lb.}$$

$$CBE = 0.5\,W_f + W_r(1 - 0.127G)$$

$$CBE = 0.5 \times 3{,}098 + 5{,}000 \times 1.833 \times (1 - 0.127 \times 0.9)$$

$$CBE = 9{,}670 \text{ lb.}$$

$$PT_{up} = (PPRL - CBE)\,S/2 = (14{,}060 - 9{,}670)\,54/2$$

$$PT_{up} = 118{,}530 \text{ lb.}$$

$$PT_{down} = (CBE - MPRL)\,S/2 = (9{,}670 - 6{,}320)\,54/2$$

$$PT_{down} = 90{,}450 \text{ lb.}$$

$$HHP = 7.36 \times 10^{-6} \times GL \times BPD$$

$$HHP = 7.36 \times 10^{-6} \times 0.9 \times 4{,}500 \times 156 = 4.7.$$

If an electric motor is to be used, then

Electric motor HP = 4.7 × 2 = 9.4 HP.

The above calculations can be used in conjunction with table 12.4 to select a pumping unit size. A review of this chart indicates that an API 160-173-64 unit will be required. In this case, the peak torque is just beyond the range of an API 114 unit.

API RP 11L Solution

The API RP 11L example problem solution is shown in figure 12.13. The form illustrated is used to make designs using the RP 11L procedure. Blank pads of the form are available from API. The following discussion refers to the form in figure 12.13. The numbered items refer to the numbered sections of the form. The

Object: To solve for—S_p, PD, PPRL, MPRL, PT, PRHP, and CBE

Known or Assumed Data:

Fluid Level, H = 4,500 ft. — Pumping Speed, N = 16 SPM — Plunger Diameter, D = 1.50 in.

Pump Depth, L = 5,000 ft. — Length of Stroke, S = 54 in. — Spec. Grav. of Fluid, G = 0.9

Tubing Size 2 in. Is it anchored? Yes, (No) — Sucker Rods 33.8 % - 7/8" & 66.2 % - 3/4"

Record Factors from Tables 4.1 & 4.2:

1. W_r = 1.833 (Table 4.1, Column 3)
2. E_r = .804 x 10⁻⁶ (Table 4.1, Column 4)
3. F_c = 1.082 (Table 4.1, Column 5)
4. E_t = .307 x 10⁻⁶ (Table 4.2, Column 5)

Calculate Non-Dimensional Variables:

5. $F_o = .340 \times G \times D^2 \times H = .340 \times 0.9 \times 2.25 \times 4500 = 3,098$ lbs.
6. $1/k_r = E_r \times L = .804 \times 10^{-6} \times 5,000 = 4.020 \times 10^{-3}$ in/lb.
7. $Sk_r = S \div 1/k_r = 54 \div 4.020 \times 10^{-3} = 13,433$ lbs.
8. $F_o/Sk_r = 3,098 \div 13,433 = .231$
9. $N/N_o = NL \div 245{,}000 = 16 \times 5,000 \div 245{,}000 = .326$
10. $N/N_o' = N/N_o \div F_c = .326 \div 1.082 = .301$
11. $1/k_t = E_t \times L = .307 \times 10^{-6} \times 5,000 = 1.535 \times 10^{-3}$ in/lb.

Solve for S_p and PD:

12. $S_p/S = .86$ (Figure 4.1)
13. $S_p = [(S_p/S) \times S] - [F_o \times 1/k_t] = [.86 \times 54] - [3,098 \times 1.535 \times 10^{-3}] = 41.7$ in.
14. $PD = 0.1166 \times S_p \times N \times D^2 = 0.1166 \times 41.7 \times 16 \times 2.25 = 175$ barrels per day

If the calculated pump displacement fails to satisfy known or anticipated requirements, appropriate adjustments must be made in the assumed data and steps 1 through 14 repeated. When the calculated pump displacement is acceptable, proceed with the Design Calculation.

Determine Non-Dimensional Parameters:

15. $W = W_r \times L = 1.833 \times 5,000 = 9,165$ lbs.
16. $W_{rf} = W[1 - (.128G)] = 9,165 [1 - (.128 \times .9)] = 8,110$ lbs.
17. $W_{rf}/Sk_r = 8,110 \div 13,433 = .604$

Record Non-Dimensional Factors from Figures 4.2 through 4.6:

18. $F_1/Sk_r = .465$ (Figure 4.2)
19. $F_2/Sk_r = .213$ (Figure 4.3)
20. $2T/S^2k_r = .37$ (Figure 4.4)
21. $F_3/Sk_r = .29$ (Figure 4.5)
22. $T_a = .997$ (Figure 4.6)

Solve for Operating Characteristics:

23. $PPRL = W_{rf} + [(F_1/Sk_r) \times Sk_r] = 8,110 + [.465 \times 13,433] = 14,356$ lbs.
24. $MPRL = W_{rf} - [(F_2/Sk_r) \times Sk_r] = 8,110 - [.22 \times 13,433] = 5,249$ lbs.
25. $PT = (2T/S^2k_r) \times Sk_r \times S/2 \times T_a = .37 \times 13,433 \times 27 \times .997 = 133,793$ lb inches
26. $PRHP = (F_3/Sk_r) \times Sk_r \times S \times N \times 2.53 \times 10^{-6} = .29 \times 13,433 \times 54 \times 16 \times 2.53 \times 10^{-6} = 8.5$
27. $CBE = 1.06(W_{rf} + 1/2\ F_o) = 1.06 \times (8,110 + 1,549) = 10,239$ lbs.

Figure 12.13. Example design calculations for conventional sucker rod pumping system. From API RP 11L, courtesy of API.

TABLE 12.5
TUBING DATA

Tubing Size	Outside Diameter, in.	Inside Diameter, in.	Metal Area, sq in.	Elastic Constant, in./lb-ft E_t
1.900	1.900	1.610	0.800	0.500×10^{-6}
2³/₈	2.375	1.995	1.304	0.307×10^{-6}
2⁷/₈	2.875	2.441	1.812	0.221×10^{-6}
3½	3.500	2.992	2.590	0.154×10^{-6}
4	4.000	3.476	3.077	0.130×10^{-6}
4½	4.500	3.958	3.601	0.111×10^{-6}

SOURCE: API RP 11L, *Design Calculations for Sucker Rod Pumping Systems*

known or assumed data are copied in the upper portion of the form.

The following data are then recorded:

1. $W_r = 1.833$ lb/ft Table 12.3
2. $E_r = 0.804 \times 10^{-6}$ Table 12.3
3. $F_c = 1.082$ Table 12.3
4. $E_t = 0.307 \times 10^{-6}$ Table 12.5

The following nondimensional variables are calculated:

5. F_o = fluid weight $= 0.34 \times 0.9 \times D^2 \times H$
   $F_o = 0.34 \times 0.9 \times (1.5)^2 \times 4{,}500$
   $= 3{,}098$ lb.

The weight of the fluid is the same as that calculated by the Mills method, since the full plunger area was used in both calculations.

6. $1/k_r = E_r \times L = 0.804 \times 10^{-6} \times 5{,}000$
   $= 4.02 \times 10^{-3}$ inch/lb.

The rods will stretch 4.02×10^{-3} or 0.00402 inch for each pound of load imposed.

7. $Sk_r = S \div 1/k_r = 54 \div 4.02 \times 10^{-3}$
   $= 13{,}433$ lb.

Sk_r is the load that would have to be imposed on the rod string to stretch it the stroke length, or 54 inches.

8. $\dfrac{F_o}{Sk_r} = 3{,}098 \div 13{,}433 = 0.231$

F_o/Sk_r is the ratio of the rod stretch to the stroke length.

9. $N/N_o = NL \div 245{,}000$
   $= 16 \times 5{,}000 \div 245{,}000$
   $= 0.326.$

10. $N/N_o' = N/N_o \div F_c = 0.326 \div 1.082$
    $= 0.301$

where N_o = spm at the natural frequency of a single rod string and N_o' = spm at the natural frequency of a tapered rod string.

11. $1/k_t = E_t \times L = 0.307 \times 10^{-6} \times 5{,}000$
    $= 1.535 \times 10^{-3}$ inch/lb.

This number indicates that the tubing will stretch 1.535×10^{-3} or 0.001535 inch for each pound of fluid load imposed.

Calculations for S_p and pump displacement, *PD,* follow.

12. $S_p/S = 0.86$. This value is obtained from figure 12.7. Enter the bottom of the chart at a value of $N/N_o' = 0.301$ and move vertically up to a value of $F_o/Sk_r = 0.231$. Move horizontally to the left and read $S_p/S = 0.86$ on the abscissa of the graph. It is necessary to interpolate between a value of $F_o/Sk_r = 0.2$ and 0.3.

13. $S_p = (S_p/S \times S) - (F_o \times 1/k_t)$
    $= (0.86 \times 54) - (3{,}098 \times 1.535 \times 10^{-3})$
    $S_p = 41.7$ inches

The rod stretch and pump overtravel are accounted for in the value of S_p/S obtained from figure 12.7. The $F_o \times 1/k_t$ value that is subtracted is for the tubing stretch, since the tubing is unanchored.

14. $PD = 0.1166 \times S_p \times N \times D^2$
    $= 0.1166 \times 41.7 \times 16 \times (1.5)^2$
    $PD = 175$ bbl/day.

This is the same equation, in a slightly different form, that was used in the Mills solution. The Mills solution used $PD = K\ S_p\ N$, where $K = 0.1484\ A_p$.

Since $A_p = \frac{\pi}{4} D^2$

$$PD = 0.1484 \times \pi/4 \times D^2 \times S_p \times N$$
$$PD = 0.1166 \times S_p \times N$$

The 175 barrels/day is at a pump volumetric efficiency of 100%. Since a pump efficiency of 85% was assumed,

$$\text{Production} = 175 \times 0.85 = 149 \text{ BPD}.$$

A desired producing rate of 150 BPD was specified in the example problem data. This value is close enough to the desired value to allow proceeding. If the value were considerably lower, then it would be necessary to assume new values for S, N, or D.

The next steps involve the calculation of some more nondimensional parameters.

15. $W = W_r \times L = 1.833 \times 5{,}000 = 9{,}165$ lb.

W is the weight of the rods obtained by multiplying the weight/ft times the length.

16. $W_{rf} = W(1 - 0.128G)$
$= 9{,}165\,(1 - 0.128 \times 0.9)$
$W_{rf} = 8{,}110$ lb.

W_{rf} is the weight of the rods in fluid. The term $W \times 0.128G$ is the buoyancy of the rods in the fluid.

17. $W_{rf}/Sk_r = 8{,}110 \div 13{,}433 = 0.604.$

This term is used in obtaining a torque adjustment factor T_a in the peak torque calculation.

The next step is to record some nondimensional variables from the charts to use in the calculations of *PPRL, MPRL, PT, PRHP,* and *CBE.*

18. $\frac{F_1}{Sk_r} = 0.465$

This value of F_1/Sk_r is obtained by entering the bottom of the chart in figure 12.8 with an $N/N_o = 0.326$. Move vertically on the $N/N_o = 0.326$ line until it intersects the point where $F_o/Sk_r = 0.231$. The value of $F_1/Sk_r = 0.465$ is then read on the abscissa. Actually the best value that can be obtained by reading the chart is 0.47, since only one place can be read and one place can be estimated. The value of 0.465 is used because this is the value used in the API example problem.

Using the value of $F_1/Sk_r = 0.465$ obtained in no. 18, the *PPRL* is in the following step.

23. $PPRL = W_{rf} + (F_1/Sk_r \times Sk_r)$
$= 8{,}110 + 0.465 \times 13{,}433$
$PPRL = 14{,}356$ lb.

W_{rf} is the weight of the rods in fluid, and $F_1/Sk_r \times Sk_r$ is the added load due to acceleration and other forces.

19. $F_2/Sk_r = 0.213$

The value of F_2/Sk_r is read from the chart in figure 12.9 in exactly the same manner used to obtain F_1/Sk_r in the preceding step.

The *MPRL* is then calculated as follows:

24. $MPRL = W_{rf} - (F_2/Sk_r \times Sk_r)$
$= 8{,}110 - 0.213 \times 13{,}433$
$MPRL = 5{,}249$ lb.

The weight of the rods in fluids less the dynamic load, $F_2/Sk_r \times Sk_r$, is equal to the *MPRL.*

20. $2T/S^2k_r = 0.37$

This value of $2T/S^2k_r$ is obtained from the chart in figure 12.10 in the same manner as that used to obtain F_1/Sk_r and F_2/Sk_r previously. Before calculating the peak torque, *PT*, it is necessary to obtain a torque adjustment factor, T_a.

22. $T_a = 0.997$

The value of T_a is obtained as follows:

Enter the bottom of the chart in figure 12.11 with a value of $N/N_o' = 0.301$. At the intersection of the $N/N_o' = 0.301$ with $F_o/Sk_r = 0.231$, read -0.1%.

$$T_a = 1 + \% \text{ value from chart} \times \frac{W_{rf} - 0.3}{0.1}$$

$$T_a = 1 + \frac{(-0.1\% \times 0.604 - 0.3)}{0.1}$$

$$T_a = 1 + (-0.3\%) = 1 - 0.003 = 0.997.$$

Using the values of $2T/S^2k_r$ and T_a, PT is calculated as follows:

25. $PT = 2T/S^2k_r \times Sk_r \times S/2 \times T_a$
$= 0.37 \times 13{,}433 \times 27 \times 0.997$
$PT = 133{,}793$ inch-lb.

The polished rod horsepower, *PRHP*, is calculated using a value of $F_3/Sk_r = 0.29$ obtained from figure 12.12.

26. $PRHP = F_3/Sk_r \times Sk_r \times S \times N \times 2.53 \times 10^{-6}$
$PRHP = 0.29 \times 13{,}433 \times 54 \times 16 \times 2.53 \times 10^{-6}$
$PRHP = 8.5.$

The counterbalance effect, *CBE*, is calculated by the formula–

27. $CBE = 1.06\,(W_{rf} + \frac{1}{2}F_o)$
$CBE = 1.06\,(8{,}110 + 3{,}098/2) = 10{,}239$ lb.

Comparison of Results–Mills and API RP 11L Calculations

A complete set of calculations has been made for both the Mills and the API RP 11L methods. Table 12.6 compares the results obtained by the two methods on the example problem.

The fact that some of the results are shown to six significant figures is not meant to suggest that the pumping unit calculations by either method is this accurate. The API data were copied exactly as they are presented in the example problem in RP 11L.

TABLE 12.6
COMPARISON OF RESULTS

	Mills	API RP 11L
Pump Displacement, BPP	184	175
Production BPD @ E_v = 85%	156	149
PPRL, lb	14,060	14,356
MPRL, lb	6,320	5,249
PT, inch-lb	118,530	133,793
PRHP	9.4	8.5
CBE, lb	9,670	10,239

The above tabulation indicates that the two methods give comparable results on this problem except for the peak torque. The Mills method generally tends to understate the peak torque.

One of the reasons that the two methods give similar results is that the Mills method was modified to compute the fluid load on the basis of the full plunger area. The fluid load in the original Mills method was based on the plunger area minus the area of the connecting sucker rod $(A_p - A_r)$.

Neither method is completely accurate, and the results must be tempered with experience. The RP 11L is probably the most widely accepted method today, but thousands of installations have been made using the Mills method, and it is still used by some companies.

GLOSSARY

A

absolute permeability *n:* a measure of the ability of a single fluid (such as water, gas, or oil) to flow through a rock formation when the formation is totally filled (saturated) with that fluid. The permeability measure of a rock filled with a single fluid is different from the permeability measure of the same rock filled with two or more fluids.

accelerator *n:* a chemical additive that reduces the setting time of cement.

acidize *v:* to treat oil-bearing limestone or other formations with acid for the purpose of increasing production. Hydrochloric or other acid is injected into the formation under pressure. The acid etches the rock, enlarging the pore spaces and passages through which the reservoir fluids flow. The acid is held under pressure for a period of time and then pumped out, after which the well is swabbed and put back into production. Chemical inhibitors combined with the acid prevent corrosion of the pipe.

anticline *n:* an arched, inverted-trough configuration of folded rock layers.

artificial lift *n:* any method used to raise oil to the surface through a well after reservoir pressure has declined to the point at which the well no longer produces by means of natural energy. Sucker rod pumps, gas lift, hydraulic pumps, and submersible electric pumps are the most common forms of artificial lift.

B

back-pressure *n:* the pressure maintained on equipment or systems through which a fluid flows.

backwash *v:* to reverse the flow of fluid from a water injection well in order to get rid of sediment that has clogged the wellbore.

beam pumping unit *n:* a machine designed specifically for sucker rod pumping, using a horizontal member (walking beam) that is worked up and down by a rotating crank to produce reciprocating motion.

blanking plug *n:* a plug used to cut off flow of liquid.

block squeeze *n:* a technique in squeeze cementing in which (1) the zone below the producing interval is perforated and a high-pressure squeeze done; (2) the zone above the producing interval is perforated and squeezed off in a similar manner; (3) the hole is drilled out; and (4) the producing interval is perforated. The purpose of block squeezing is to isolate the producing interval and prevent communication with the sand immedicately above and below the producing interval, but the technique is very questionable.

bottomhole pump *n:* any of the rod pumps, high-pressure liquid pumps, or centrifugal pumps located at or near the bottom of the well and used to lift the well fluids.

bottom water *n:* water found below oil and gas in a producing formation.

bottom-water drive *n:* See *water drive.*

bradenhead squeezing *n:* the process by which hydraulic pressure is applied to a well to force fluid or cement outside the wellbore without the use of a packer. The bradenhead, or casinghead, is closed to shut off the annulus when making a bradenhead squeeze. Although this term is still used, the term *bradenhead* is obsolete.

brake horsepower *n:* the power produced by an engine as it is measured by the force applied to a friction brake or by an absorption dynamometer applied to the shaft or the flywheel.

bridge plug *n:* a downhole tool, composed primarily of slips, a plug mandrel, and a rubber sealing element, that is run and set in casing to isolate a lower zone while an upper section is being tested or cemented.

bubble point *n:* 1. the temperature and pressure at which part of a liquid begins to convert to gas. For example, if a certain volume of liquid is held at constant pressure, but its temperature is increased, a point is reached when bubbles of gas begin to form in the liquid. That is the bubble point. Similarly, if a certain volume of liquid is held at a constant temperature but the pressure is reduced, the point at which gas begins to form is the bubble point. 2. the temperature and pressure at which gas, held in solution in crude oil, breaks out of solution as free gas.

bullheading *n:* overpowering a well by pumping a kill fluid down the tubing or casing and killing the well.

C

caliper log *n:* a record showing variations in wellbore diameter by depth, indicating undue enlargement due to caving in, washout, or other causes. The caliper log also reveals corrosion, scaling, or pitting inside tubular goods.

cement bond survey *n:* an acoustic survey or sonic-logging method that records the quality or hardness of the cement used in the annulus to bond the casing and the formation. Casing that is well bonded to the formation transmits an acoustic signal quickly; poorly bonded casing transmits a signal slowly.

cementing *n:* the application of a liquid slurry of cement and water to various points inside or outside the casing. Types of cementing are primary, secondary, and squeeze.

centipoise *n:* one-hundredth of a poise.

centrifugal pump *n:* a pump with an impeller or rotor, an impeller shaft, and a casing, which discharges fluid by centrifugal force.

circulation sleeve *n:* See *sliding sleeve.*

circulation squeeze *n:* a variation of squeeze cementing for wells with two producing zones in which (1) the upper fluid sand is perforated; (2) tubing is run with a packer, and the packer is set between the two perforated intervals; (3) water is circulated between the two zones to remove as much mud as possible from the channel; (4) cement is pumped through the channel and circulated; (5) the packer is released and picked up above the upper perforation, a low squeeze pressure is applied, and the excess cement is circulated out. The process is applicable where there is communication behind the pipe between the two producing zones because of channeling of the primary cement or where there is essentially no cement in the annulus.

coefficient of expansion *n:* the increment in volume of a unit volume of solid, liquid, or gas for a rise of temperature of 1° at constant pressure. Also called coefficient of cubical expansion, coefficient of thermal expansion, expansion coefficient, expansivity.

coiled-tubing workover *n:* a workover performed with a continuous steel tube, normally 3/4" to 1" OD, which is run into the well in one piece inside the normal tubing. Lengths of the tubing up to 16,000 feet long are stored on the surface on a reel in a manner similar to that used for wireline. The unit is rigged up over the wellhead. The tubing is injected through a control head that seals off the tubing and makes a pressure-tight connection. A unique feature of the unit is that it allows continuous circulation during its lowering into the hole.

competitive field *n:* an oil or gas field comprised of wells operated by various operators.

complete a well *v:* to finish work on a well and bring it to productive status. See *well completion.*

compressive strength *n:* the degree of resistance of a material to a force acting along one of its axes in a manner tending to collapse it; usually expressed in pounds of force per square inch (psi) of surface affected.

concentric-tubing workover *n:* a workover performed with a small-diameter tubing work string inside the normal tubing. Equipment needed is essentially the same as that for a conventional workover except that it is smaller and lighter.

condensate *n:* a light hydrocarbon liquid obtained by condensation of hydrocarbon vapors. It consists of varying proportions of butane, propane, pentane, and heavier fractions, with little or no methane or ethane.

coning *n:* the enroachment of reservoir water or gas into the oil column and well because of production.

connate water *n:* water retained in the pore spaces, or interstices, of a formation from the time the formation was created.

core *n:* a cylindrical sample taken from a formation for geological analysis. Usually a conventional core barrel is substituted for the bit and procures a sample as it penetrates the formation. *v:* to obtain a solid, cylindrical formation sample for analysis.

counterbalance effect *n:* the effect of counterweights on a beam pumping system. The approximate ideal counterbalance effect is equal to half the weight of the fluid plus the buoyant weight of the rods.

counterbalance weight *n:* a weight applied to compensate for existing weight or force. On pumping units in oil production, counterweights are used to offset the weight of the column of sucker rods and fluid on the upstroke of the pump, and the weight of the rods on the downstroke.

crank arm *n:* a steel member connected to each end of the shaft extending from each side of the speed reducer on a beam pumping unit.

crossover packer *n:* a type of packer developed for a dual-completion well in which there are both an oil and a gas zone, with the gas zone on the bottom.

D

darcy *n:* a unit of measure of permeability. A porous medium has a permeability of 1 darcy when a differential pressure drop of 1 atmosphere across a sample 1 cm long and 1 cm^2 in cross section will force a liquid of 1-cp viscosity through the sample at the rate of 1 cm^3 per second. The permeability of reservoir rocks is usually so low that it is measured in millidarcy units.

differential pressure *n:* the difference between two fluid pressures; for example, the difference between the pressure in a reservoir and in a wellbore drilled in the reservoir, or between atmospheric pressure at sea level and at 10,000 feet. Also called pressure differential.

dip *n:* also called formation dip. See *formation dip.*

dissolved-gas drive *n:* a solution-gas drive. See *reservoir drive mechanism.*

drawdown *n:* 1. the difference between static and flowing bottomhole pressures. 2. the distance between the static level and the pumping level of the fluid in the annulus of a pumping well.

dump bailer *n:* a bailing device with a release valve, usually of the disk or flapper type, used to place or spot material (such as cement lurry) at the bottom of the well.

E

edgewater *n:* the water that touches the edge of the oil in the lower horizon of a formation.

edgewater drive *n:* See *water drive.*

electric submersible pumping *n:* a form of artificial lift that utilizes an electric submersible multistage centrifugal pump. Electric power is conducted to the pump by a cable attached to the tubing.

F

filter cake *n:* the layer of concentrated solids from the drilling mud or cement slurry that forms on the walls of the borehole opposite permeable formations; also called wall cake or mud cake.

fish *n:* an object that is left in the wellbore during drilling or workover operations and that must be recovered before work can proceed. It can be anything from a piece of scrap metal to a part of the drill stem. *v:* 1. to recover from a well any equipment left there during drilling operations, such as a lost bit or drill collar or part of the drill string. 2. to remove from an older well certain pieces of equipment (such as packers, liners, or screen pipe) to allow reconditioning of the well.

flowing bottomhole pressure *n:* pressure at the bottom of the wellbore during normal oil production.

fluid loss *n:* the undesirable migration of the liquid part of the drilling mud or cement slurry into a formation, often minimized or prevented by the blending of additives with the mud or cement.

formation damage *n:* the reduction of permeability in a reservoir rock caused by the invasion of drilling fluid and treating fluids to the section adjacent to the wellbore. It is often called skin damage. See *skin.*

formation dip *n:* the angle at which a formation bed inclines away from the horizontal. *Dip* is also used to describe the orientation of a fault.

formation fracturing *n:* a method of stimulating production by increasing the permeability of the producing formation. Under extremely high hydraulic pressure, a fluid (such as water, oil, alcohol, dilute hydrochloric acid, liquefied petroleum gas, or foam) is pumped downward through tubing or drill pipe and forced into the perforations in the casing. The fluid enters the formation and parts or fractures it. Sand grains, aluminum pellets, glass beads, or similar materials are carried in suspension by the fluid into the fractures. These are called propping agents or proppants. When the pressure is released at the surface, the fracturing fluid returns to the well, and the fractures partially close on the proppants, leaving channels for oil to flow through them to the well. This process is often called a frac job.

formation volume factor *n:* the factor that is used to convert stock tank barrels of oil to reservoir barrels. It is the ratio between the space occupied by a barrel of oil containing solution gas at reservoir conditions and a barrel of dead oil at surface conditions. Also called reservoir volume factor.

fracturing *n:* shortened form of formation fracturing. See *formation fracturing.*

G

gas cap *n:* a free-gas phase overlying an oil zone and occurring within the same producing formation as the oil.

gas-cap drive *n:* drive energy supplied naturally (as a reservoir is produced) by the expansion of gas in a cap overlying the oil in the reservoir. See *reservoir drive mechanism.*

gas lift *n:* the process of raising or lifting fluid from a well by injecting gas down the well through tubing or through the tubing-casing annulus. Injected gas aerates the fluid to make it exert less pressure than the formation does; consequently, the higher formation pressure forces the fluid out of the wellbore. Gas may be injected continuously or intermittently, depending on the producing characteristics of the well and the arrangement of the gas-lift equipment.

gear reduction unit *n:* a gear train that lowers the output speed.

gravel packing *n:* a method of well completion in which a slotted or perforated liner is placed in the well and surrounded by small-sized gravel. The well is sometimes enlarged by underreaming at the point where the gravel is packed. The mass of gravel excludes sand from intruding in the well but allows continued rapid production.

H

horsehead *n:* the curved section of the walking beam of a beam pumping unit, which is located on the oilwell end and from which the bridle is suspended.

hydraulic fracturing *n:* an operation in which a specially blended liquid is pumped down a well and into a formation under pressure high enough to cause the formation to crack open, forming passages through which oil can flow into the wellbore. See *formation fracturing.*

hydraulic horsepower *n:* a measure of the power of a fluid under pressure.

hydraulic pump *n:* a device that lifts oil from wells without the use of sucker rods. See *hydraulic pumping.*

hydraulic pumping *n:* a method of pumping oil from wells by using a downhole pump without sucker rods. Subsurface hydraulic pumps consist of two reciprocating pumps coupled and placed in the well. One pump functions as an engine and drives the other pump (the production pump). Surface power is supplied from a standard engine-driven pump. The downhole engine is usually operated by clean crude oil under pressure (power oil) that is drawn from a power-oil or settling tank by a triplex plunger pump. If a single string of tubing is used, power oil is pumped down the tubing string to the pump, which is seated in the string, and a mixture of power oil and produced fluid is returned through the casing-tubing annulus. If two parallel strings are used, one supplies the power oil to the pump while the other returns the exhaust and the produced oil to the surface. The hydraulic pump may be used to pump several wells from a central source and has been used to lift oil from depths of more than 10,000 feet.

hydrostatic head *n:* also called hydrostatic pressure. See *hydrostatic pressure.*

hydrostatic pressure *n:* the force exerted by a body of fluid at rest, which increases directly with the density and the depth of the fluid and is expressed in psi. The hydrostatic pressure of fresh water is 0.433 psi per foot of depth. In drilling, the term refers to the pressure exerted by the drilling fluid in the wellbore. In a water drive field, the term refers to the pressure that may furnish the primary energy for production.

I

inflow performance relationship *n:* the relation between the midpoint pressure of the producing interval and the liquid inflow rate of a producing well.

injection well *n:* a well through which fluids are injected into an underground stratum usually to increase reservoir pressure and to displace oil.

intermediate casing string *n:* the string of casing set in a well after the surface casing is set to keep the hole from caving and to seal off troublesome formations. The string is sometimes called protection casing.

isopach map *n:* a geological map of subsurface strata showing the various thicknesses of a given formation underlying an area. It is widely used in calculating reserves and in planning secondary recovery projects.

J

jet-perforate *v:* to create a hole through the casing with a shaped charge of high explosives instead of a gun that fires projectiles. The loaded charges are lowered into the hole to the desired depth. Once detonated, the charges emit short, penetrating jets of high-velocity gases that cut holes in the casing and cement and some distance into the formation. Formation fluids then flow into the wellbore through these perforations.

K

kill *v:* 1. in drilling, to prevent a threatened blowout by taking suitable preventive measures (e.g., to shut in the well with the blowout preventers, circulate the kick out, and increase the weight of the drilling mud). 2. in production, to stop a well from producing oil and gas so that reconditioning of the well can proceed. Production is stopped by circulating a killing fluid into the hole.

L

landing nipple *n:* a device machined internally to receive the removable locking devices used to position, lock, and seal subsurface production controls in tubing. A landing nipple provides a seat at a known depth, into which various types of retrievable flow control equipment can be set.

latch sub *n:* a device, usually with segmented threads, run with seal subs on the bottom of a tubing string and latched into a permanent packer to prevent tubing movement.

locator sub *n:* a device, larger than the bore of a permanent packer, which is run with seal subs on the bottom of a tubing string and used to locate the top of a permanent packer.

log cross section *n:* a cross section of a reservoir or part of a reservoir constructed with electric or radioactive logs.

lost circulation *n:* the quantities of whole mud lost to a formation, usually in cavernous, fissured, or coarsely permeable beds, evidenced by the complete or partial failure of the mud to return to the surface as it is being circulated in the hole. Lost circulation can lead to a blowout and, in general, reduce the efficiency of the drilling operation. Also called lost returns.

lubricator *n:* a specially fabricated length of casing or tubing usually placed temporarily above a valve on top of the casinghead or tubing head; used to run swabbing or perforating tools into a producing well; provides a method for sealing off pressure and thus should be rated for highest anticipated pressure.

M

modulus of elasticity *n:* the ratio of the increment of some specified form of stress to the increment of some specified form of strain, such as Young's modulus, bulk modulus, or shear modulus. Also called coefficient of elasticity, elasticity modulus, elastic modulus.

moment of inertia *n:* the sum of the products formed by multiplying the mass (or sometimes, the area) of each element of a figure by the square of its distance from a specified line. Also called rotational inertia.

multiple completion *n:* an arrangement for producing a well in which one wellbore penetrates two or more petroleum-bearing formations. In one type, multiple tubing strings are suspended side by side in the production casing string, each a different length and each packed off to prevent the commingling of different reservoir fluids. Each reservoir is then produced through its own tubing string. Alternately, a small-diameter production casing string may be provided for each reservoir, as in multiple miniaturized or multiple tubingless completions.

N

Newtonian fluid *n:* a simple fluid in which the state of stress at any point is proportional to the time rate of strain at that point; the proportionality factor is the viscosity coefficient.

O

offset well *n:* a well drilled on a tract of land next to another owner's tract on which there is a producing well.

open-hole completion *n:* a method of preparing a well for production in which no production casing or liner is set opposite the producing formation. Reservoir fluids flow unrestricted into the open wellbore. An open-hole completion has limited use in rather special situations. Also called a barefoot completion.

P

packer *n:* a piece of downhole equipment, consisting of a sealing device, a holding or setting device, and an inside passage for fluids, used to block the flow of fluids through the annular space between the tubing and the wall of the wellbore by sealing off the space between them. It is usually made up in the tubing string some distance above the producing zone. A packing element expands to prevent fluid flow except through the inside bore of the packer and into the tubing. Packers are classified according to configuration, use, and method of setting and whether or not they are retrievable (that is, whether they can be removed when necessary, or whether they must be milled or drilled out and thus destroyed).

pay *n:* See *pay sand.*

pay sand *n:* the producing formation, often one that is not even sandstone. It is also called pay, pay zone, and producing zone.

pay zone *n:* See *pay sand.*

perforated completion *n:* 1. a well completion in which the producing zone or zones are cased through, cemented, and perforated to allow fluid flow into the wellbore. 2. a well completed by this method.

perforating gun *n:* a device, fitted with shaped charges or bullets, that is lowered to the desired depth in a well and fired to create penetrating holes in casing, cement, and formation.

perforation *n:* a hole made in the casing, cement, and formation, through which formation fluids enter a wellbore. Usually several perforations are made at a time.

perforation washer *n:* a device utilizing rubber cups run on the tubing string and used to wash the perforations of wells completed in unconsolidated sands with water.

permeability *n:* 1. a measure of the ease with which a fluid flows through the connecting pore spaces of rock or cement. The unit of measurement is the millidarcy. 2. fluid conductivity of a porous medium. 3. ability of a fluid to flow within the interconnected pore network of a porous medium.

pinchout *n:* a geological structure that forms a trap for oil and gas when a porous and permeable rock ends at or stops against an impervious formation.

pitman *n:* the arm that connects the crank to the walking beam on a pumping unit by means of which rotary motion is converted to reciprocating motion.

plug back *v:* to place cement in or near the bottom of a well to exclude bottom water, to sidetrack, or to produce from a formation already drilled through. Plugging back can also be accomplished with a mechanical plug set by wireline, tubing, or drill pipe.

plunger overtravel *n:* an increase in the effective stroke length of the plunger of a sucker rod pump, caused by the elongation of the rod string due to the dynamic loads imposed by the pumping cycle.

poise *n:* the viscosity of a liquid in which a force of 1 dyne (a unit of measurement of small amounts of force) exerted tangentially on a surface of 1 cm^2 of either of two parallel planes 1 cm apart will move one plane at the rate of 1 cm per second in reference to the other plane, with the space between the two planes filled with the liquid.

polished rod *n:* the topmost portion of a string of sucker rods, used for lifting fluid by the rod-pumping method. It has a uniform diameter and is smoothly polished to effectively seal pressure in the stuffing box attached to the top of the well.

portland cement *n:* the cement most widely used in oilwells. It is made from raw materials such as limestone, clay or shale, and iron ore.

positive-displacement pump *n:* a pump that moves a measured quantity of liquid with each stroke of a piston or each revolution of vanes or gears; a reciprocating pump or a rotary pump.

pozzolan *n:* a natural or artificial siliceous material commonly added to portland cement mixtures to impart certain desirable properties. Added to oilwell cements, pozzolans reduce slurry weight and viscosity, increase resistance to sulfate attack, and influence factors such as pumping time, ultimate strength, and watertightness.

pressure differential *n:* also called differential pressure. See *differential pressure.*

pressure sink *n:* a condition in which the pressure at the wellbore is less than the reservoir pressure. Flow of oil from the reservoir to the wellbore occurs because of this pressure differential.

primary recovery *n:* the first stage of oil production in which natural reservoir drives are used to recover oil.

prime mover *n:* an internal-combustion engine or a turbine that is the source of power for driving a machine or machines.

production casing *n:* the last string of casing that is set in a well, inside of which is usually suspended a tubing string.

productivity index *n:* a well-test measurement indicative of the amount of oil or gas a well is capable of producing. It may be expressed as

$$PI = \frac{q}{p_e - p_w}$$

where

PI = productivity index (bbl/d or Mcf/d per psi of pressure differential)

q = rate of production (bbl/d or Mcf/d)

p_e = static bottomhole pressure (psi)

p_w = flowing bottomhole pressure (psi)

psia *abbr:* pounds per square inch absolute. Psia is equal to the gauge pressure plus the pressure of the atmosphere at that point.

psig *abbr:* pounds per square inch gauge.

R

radial flow *n:* the flow pattern of fluids flowing into a wellbore from the surrounding drainage area.

relative permeability *n:* a measure of the ability of two or more fluids (such as water, gas, and oil) to flow through a rock formation when the formation is totally filled with several fluids. The permeability measure of a rock filled with two or more fluids is different from the permeability measure of the same rock filled with only a single fluid.

relief *n:* the elevations or inequalities of a land surface.

reservoir *n:* a subsurface, porous, permeable rock body in which oil and/or gas has accumulated. Most reservoir rocks are limestones, dolomites, sandstones, or a combination of these. The three basic types of hydrocarbon reservoirs are oil, gas, and condensate. An oil reservoir generally contains three fluids–gas, oil, and water–with oil the dominant product. In the typical oil reservoir, these fluids become vertically segregated because of their different densities. Gas, the lightest, occupies the upper part of the reservoir rocks; water, the lower part; and oil, the intermediate section. In addition to its occurrence as a cap or in solution, gas may accumulate independently of the oil; if so, the reservoir is called a gas reservoir. Associated with the gas, in most instances, are salt water and some oil. In a condensate reservoir, the hydrocarbons may exist as a gas, but, when brought to the surface, some of the heavier ones condense to a liquid.

reservoir drive mechanism *n:* the process in which reservoir fluids are caused to flow out of the reservoir rock and into a wellbore by natural energy. Gas drives depend on the fact that, as the reservoir is produced, pressure is reduced, allowing the gas to expand and provide the principal driving energy. Water drive reservoirs depend on water and rock expansion to force the hydrocarbons out of the reservoir and into the wellbore.

reservoir volume factor *n:* See *formation volume factor.*

retarder *n:* a substance added to cement to prolong the setting time so that the cement can be pumped into place. Retarders are used for cementing in high-temperature formations.

S

seal sub *n:* a smoothly finished steel tube with rubber or synthetic seal rings run on the bottom of the tubing string and seated in a permanent packer in order to make a pressure seal.

secondary recovery *n:* 1. the use of waterflooding or gas injection to maintain formation pressure during primary production and to reduce the rate of decline of the original reservoir drive. 2. waterflooding of a depleted reservoir. 3. the first improved recovery method of any type applied to a reservoir to produce oil not recoverable by primary recovery methods.

shut-in bottomhole pressure *n:* the pressure at the bottom of a well when the surface valves on the well are completely closed, caused by formation fluids at the bottom of the well.

skin *n:* 1. the area of the formation that is damaged because of the invasion of foreign substances into the exposed section of the formation adjacent to the wellbore during drilling and completion. 2. the pressure drop from the outer limits of drainage to the wellbore caused by the relatively thin veneer (or skin) of the affected formation. Skin is expressed in dimensionless units; a positive value denotes formation damage, and a negative value indicates improvement.

sliding sleeve *n:* a special device placed in a string of tubing, which can be operated by a wireline tool to open or close orifices to permit circulation between the tubing and the annulus. It may also be used to open or shut off production from various intervals in a well. Also called sliding-sleeve nipple or circulation sleeve.

slurry *n:* a plastic mixture of cement and water that is pumped into a well to harden; there it supports the casing and provides a seal in the wellbore to prevent migration of underground fluids.

snub *v:* to force pipe or tools into a high-pressure well that has not been killed (i.e., to run pipe or tools into the well against pressure when the weights of pipe are not great enough to force the pipe through the BOPs). Snubbing usually requires an array of wireline blocks and wire rope that forces the pipe or tools into the well through a stripper head or blowout preventer until the weight of the string is sufficient to overcome the lifting effect of the well pressure on the pipe in the stripper. In workover operations, snubbing is usually accomplished by using hydraulic power to force the pipe through the stripping head or blowout preventer.

spud *v:* 1. to move the drill stem up and down in the hole over a short distance without rotation. Careless execution of this operation creates pressure surges that can cause a formation to break down, resulting in lost circulation. 2. to force a wireline tool or tubing down the hole by using a reciprocating motion.

squeeze cementing *n:* the forcing of cement slurry by pressure to specified points in a well to cause seals at the points of squeeze. It is a secondary cementing method that is used to isolate a producing formation, seal off water, repair casing leaks, and so forth.

standoff *n:* in perforating, the distance a jet or bullet must travel in the wellbore before encountering the wall of the hole.

standoff problem *n:* a problem of obtaining proper penetration with a tubing perforating gun in casing due to the casing gun's lying against one side of the casing because of hole deviation.

static bottomhole pressure *n:* pressure at the bottom of the wellbore when there is no flow of oil.

static pressure *n:* the pressure exerted by a fluid upon a surface that is at rest in relation to the fluid.

stimulation *n:* any process undertaken to enlarge oil channels or create new ones in the producing formation of a well (e.g., acidizing or formation fracturing).

structure map *n:* a map of critical horizons – subsurface oil-producing zones – in a given area.

sucker rod *n:* a special steel pumping rod. Several rods screwed together make up the mechanical link from the beam pumping unit on the surface to the sucker rod pump at the bottom of a well. Sucker rods are threaded on each end and manufactured to dimension standards and metal specifications set by the petroluem industry. Lengths are 25 or 30 feet; diameter varies from ½ to 1⅛ inches. There is also a continuous sucker rod (trade name: Corod).

surface casing *n:* also called surface pipe. See *surface pipe.*

surface pipe *n:* the first string of casing (after the conductor pipe) that is set in a well, varying in length from a few hundred to several thousand feet. Some states require a minimum length to protect freshwater sands.

swab *n:* a hollow, rubber-faced cylinder mounted on a hollow mandrel with a pin joint on the upper end to connect to the swab line. A check valve that opens upward on the lower end provides a way to remove the fluid from the well when pressure is insufficient to support flow. v: to operate a swab on a wireline to bring well fluids to the surface when the well does not flow naturally. Swabbing is a temporary operation to determine whether or not the well can be made to flow. If the well does not flow after being swabbed, a pump is installed as a permanent lifting device to bring the oil to the surface.

sweep efficiency *n:* the efficiency with which water displaces oil or gas in a water drive oil or gas field. Water flowing in from the aquifer does not displace the oil or gas uniformly but channels past certain areas due to variations in porosity and permeability.

T

torque *n:* the turning force that is applied to a shaft or other rotary mechanism to cause it to rotate or tend to do so. Torque is measured in units of length and force (foot-pounds, newton-metres).

tubingless completion *n:* a method of producing a well in which only small-diameter production casing is set through the pay zone, with no tubing or inner production string used to bring formation fluids to the surface. This type of completion has its best application in low-pressure, dry-gas reservoirs.

U

underream *v:* to enlarge the wellbore below the casing.

unloading a well *n:* removing fluid from the tubing in a well, often by means of a swab, in order to lower the bottomhole pressure in the wellbore at the perforations and induce the well to flow.

V

viscosity *n:* a measure of the resistance of a liquid to flow. Resistance is brought about by the internal friction resulting from the combined effects of cohesion and adhesion. The viscosity of petroleum products is commonly expressed in terms of the time required for a specific volume of the liquid to flow through an orifice of a specific size.

W

walking beam *n:* the horizontal steel member of a beam pumping unit, having rocking or reciprocating motion.

water block *n:* a reduction in the permeability of a formation, caused by the invasion of water into the pores.

water drive *n:* the reservoir drive mechanism in which oil is produced by the expansion of the underlying water and rock, which forces the oil into the wellbore. In general, there are two types of water drive: bottom-water drive, in which the oil is totally underlain by water, and edgewater drive, in which only the edge of the oil is in contact with the water.

waterflooding *n:* a method of enhanced oil recovery in which water is injected into a reservoir to remove additional quantities of oil that have been left behind after primary recovery. Usually, waterflooding involves the injection of water through wells specially set up for water injection and the removal of water and oil from production wells drilled adjacent to the injection wells.

well completion *n:* the activities and methods necessary to prepare a well for the production of oil and gas; the method by which a flow line for hydrocarbons is established between the reservoir and the surface. The method of well completion used by the operator depends on the individual characteristics of the producing formation or formations. Such techniques include open-hole completions, sand-exclusion completions, tubingless completions, and miniaturized completions.

wellhead *n:* the equipment installed at the surface of the wellbore. A wellhead includes such equipment as the casinghead and tubing head. adj: pertaining to the wellhead (e.g., wellhead pressure).

well stimulation *n:* See *stimulation.*

wireline workover *n:* a workover performed with wireline lowered into the well through the tubing string. A lubricator is rigged up over the wellhead and wireline tools inserted through the lubricator under pressure. A variety of wireline tools is available.

workover *n:* the performance of one or more of a variety of remedial operations on a producing oilwell to try to increase production. Examples of workover jobs are deepening, plugging back, pulling and resetting liners, squeeze cementing, and so forth.